Lerma Valero

Scientific Injection Molding Tools

José R. Lerma Valero

Scientific Injection Molding Tools

HANSER

Print-ISBN: 978-1-56990-923-2
E-Book-ISBN: 978-1-56990-942-3

Bibliographic information of the German National Library:
The German National Library lists this publication in the German National Bibliography; detailed bibliographic data are available on the Internet at http://dnb.d-nb.de.

© 2025 Carl Hanser Verlag GmbH & Co. KG, Munich
Vilshofener Straße 10 | 81679 Munich | info@hanser.de
www.hanserpublications.com
www.hanser-fachbuch.de
Editor: Dr. Mark Smith
Production Management: Cornelia Speckmaier
Cover concept: Marc Müller-Bremer, *www.rebranding.de*, Munich
Cover design: Max Kostopoulos
Cover picture: © José R. Lerma Valero, background: firefly.adobe.com
Typesetting: Eberl & Koesel Studio, Kempten

Acknowledgments

When you start a project like writing a book, you have no idea how much time it will take. Most of that time is going to be "stolen" from free time or family time.

I have to start by thanking my awesome wife Lola, my dear son Kevin, my dear daughter-in-law Ala and, of course, my little granddaughter Marina, who has stolen all of our hearts. Thanks a lot to all of them for their understanding and enormous patience during this long "second book time". I apologize for the immense quantity of "family hours" that I have dedicated to this project.

Thank you so much Lola, Kevin, Ala and Marina.

My thanks also to my parents Ildefonso and Ana Maria and my brother Juan, who unfortunately cannot see this book, but who would surely be proud of it.

To write a book, first you have to be convinced that you have something interesting to share and write about and that there are readers interested in it. Writing a second book is a hard and demanding project. In my opinion, it poses an even greater challenge, since you have a benchmark to improve upon – your previous book.

At this stage, it is important to have some collaboration and support of the people you trust, not only to support you but also, at times, to have you a sincere opinion about the project.

For this reason, it is only fair to thank people or entities that in different ways have collaborated or given their support to achieve this goal. I give special thanks to my Biesterfeld colleagues Enric Garcia, Kevin Lerma and Lidia Jimenez.

Special thanks to Ala Faller for his collaboration in the English translation of the first sample chapter in the middle of the COVID pandemic. Sincere thanks to my first teachers in the world of plastics, Juan Calero, Manuel Calero, Julio Urbano, Antonio Lorente, Jose M. Perez and Victor Ariño.

Thanks to the company in charge of the initial technical translation, Sistemes d'edició 1990 S. L.

Thanks to Albert Planas, the cover designer of the original book.

Thanks to Denis Fecci for his interesting seminars at RJG, Arinthod.

Thanks to Mark Smith Julia Diaz for their support and opinion when I explained the initial idea for the book project to them at the 2019 K fair in Düsseldorf.

Thanks to the technical centers where I develop technical training in plastic injection molding for supporting me and giving me the push I need: Aimplas, Eurecat, Aitiip, F. E. Soler, Andaltec.

Thanks to each and every one of the technicians that have attended my seminars.

Thanks to the companies, owners, managers, collaborators and teachers whom I have worked with and learned from during these more than 40 years in the world of plastics.

Thanks to the Hanser team who collaborated on this book edition, especially to Julia Diaz for her collaboration and suggestions for improvement, and to Mark Smith for his interest and follow-up on the project after our first book project meeting at 2022 K fair in Dusseldorf, and for believing in this second book edition from the beginning. Thanks a lot to Ray Brown and Rebecca Wehrmann for their collaboration in the final edition and the cover definition.

Thanks to the book sponsor companies, Biesterfeld, Arburg, Wittmann Battenfeld, Aimplas, Coscollola (KraussMaffei) for their support and patience.

Preface

The plastics injection molding industry has been evolving rapidly in many aspects, such as machinery, molds, materials, peripherals, applications, and so also has the demand for component manufacturing, but this level of evolution has not always been matched by the same level of evolution in process engineering, process mastery and parameter definition.

The definition of robust and consistent processes, which manufacture with productivity and precision in a repetitive way, has become a challenge for converters and crucial for the future. There is no doubt that those converters which master the process in a consistent way in the most dynamic and demanding sectors of our economy will survive.

The author in this book proposes that this definition of robust and consistent processes is achieved through the knowledge and application of the Scientific Injection Molding methodology. For this purpose, training of technicians and employees of the functions involved in this process definition is essential.

This book describes the preliminary steps to take into account in the implementation of this methodology, how to know our machines and especially the state of our machines through on-site tests, how to know the polymers we are going to process, the necessary equipment for the application of this methodology in injection molding plants and all the tests and tools that Scientific Injection Molding proposes, explained in detail to be able to apply them.

How to correctly perform mold validation, to establish the limits of the process window through the design of experiments, and to transfer processes from one machine to another ensuring their repeatability are all presented as fundamental tools of advanced plastic injection molding for the future.

All this is possible through the application of this methodology and through the on-machine tests that it proposes.

Based on more than 30 years of experience in the plastics injection molding industry, this book presents the steps and key points to follow for correctly implementing Scientific Molding in plastic injection molding plants and preventing common mistakes in its implementation.

It is an essential work for the improvement and mastery of the plastic injection molding process, as well as for those who want to implement an optimized, robust and consistent injection molding process definition methodology.

It is also essential for those who use and value positively the previous book by the author, "Plastics Injection Molding"; this second book is complementary and a step forward in the knowledge, definition and mastery of plastic injection molding.

Dedicated:

To my wife Lola, my son Kevin, his wife Ala and my dearest newly arrived granddaughter Marina Faller Lerma

The Author

Jose R. Lerma was born in Barcelona, Cataluña, 62 years ago. He is married, has a son, and a granddaughter.

He has a higher degree in mechanics, specializing in molds, and a master's degree in business management.

He started his professional life as trainee in a small injection molding plant.

Jose R. Lerma has dedicated close to 45 years of his professional life to the world of thermoplastics.

Most of his career, 30 years and more, has been spent in plastic injection plants, producing parts for the automotive sector, molding both technical and aesthetic parts, painting, chrome plating, etc.

The functions and responsibilities he enjoyed in these molding plants include those of Processing Engineer, Technical Department Manager, Maintenance Manager, Production Manager and Plant Manager.

Currently and for almost 16 years now, he is Technical Manager for Biesterfeld Iberica in Spain and Portugal, a leading company in polymer distribution in Europe with a portfolio of materials from the world's leading manufacturers.

Jose R. Lerma has been collaborating for more than 20 years with different technical centers in Spain as a teacher of various seminars related to plastics and injection molding, and has trained thousands of technicians in this technology and in the methodology of Scientific Injection Molding.

In 2013, he self-published the book "Advanced Manual of Thermoplastic Transformation" in Spanish, which proved highly successful among plastic injection molding technicians.

For 10 years, he has developed and successfully taught a specific seminar about Scientific Injection Molding Methodology in Spain, Portugal and some Latin American countries.

In 2020, the prestigious publisher of technical books Hanser Publications, edited and published the English version of the book, entitled "Plastics Injection Molding", which went on to become a bestseller worldwide.

In 2020, Jose R. Lerma self-published the advanced injection molding book "Scientific Injection Molding Tools" in Spanish, where he delves deeper into the methodology of Scientific Injection Molding, its key points and implementation in injection molding plants.

This second book promotes a new seminar, which is highly valued by technicians in the sector. The "Scientific Injection Molding Tools" seminar is held both in technical centers and in house in injection molding plants

Since 2016, Jose R. Lerma has written a monthly article for the plastics sector technical magazine "Plasticos Universales" from the Interempresas publishing company, with more than 70 articles appearing in this widely read magazine.

All this accumulated experience in real day-to-day cases in factories as well as the training received and the experience of providing training in seminars to technicians are reflected and shared in this book.

Contents

10   Melt Preparation

1

Scientific Injection Molding – Advanced Steps toward Implementation

1.1 Introduction

Before venturing into the implementation of any new system or methodology, it is wise and highly recommended to understand what we are going to implement and what advantages it will bring us in comparison with the system in place at the time. That is why this book begins by trying to explain what scientific injection molding is and what benefits it brings to injection molding plants. This is to encourage readers to change the way they define processes in the injection molding plant as well as in the correction of process deviations.

What is the purpose of this? The main objective is to master the process, improving it where possible, and to define repetitive, robust, cost-effective processes. The goal of any company is, on the one hand, to generate profits to compensate for the risks taken by shareholders and, on the other hand, to ensure its own survival, reinvestment, and growth.

In plastic injection molding, this objective is only achievable if we define robust processes. For example, we cannot afford to let a process be outsourced to low-cost countries because of the risk of compromising standards. Injection molding plants will experience many problems when implementing Industry 4.0 (Smart Factory) technology and methods if they do not ensure consistent, robust processes.

Understanding what this methodology is based on, the tools it proposes, and the benefits it brings is the first step in making the decision to apply it.

Scientific Injection Molding – Improving Knowledge

Many authors, economists, and intellectuals consider the new theory of economic growth to be based on improvements to people's knowledge. For some authors, the new "knowledge economy" replaces the "real economy's" production and innovation styles. In the "knowledge economy", human capital will be the main asset, so people's knowledge will be essential for that country or region's economic growth.

Therefore, it is impossible to implement this economic growth through knowledge without training ALL those involved beforehand, improving skills, gaining experience, and improving human capital (knowledge and training: the tools of the future). An economy based on knowledge and learning is a system in which the engine of value creation and/or benefits is knowledge and the capacity to build it through learning. Learning and training are essential to this economy.

We must therefore rely on specialized training centers, on experienced consultants as levers for training, and on knowledge improvement as an indispensable step before starting the project to implement the scientific injection molding system. This training will not only serve as knowledge for the successful implementation of the methodology, but it will also serve as excellent motivation for staff in respect of new knowledge, new improvements, and new tools to use that will help them understand why certain things happen in the injection molding process and how they can be mastered.

"An investment in knowledge always pays the best interest" is a phrase attributed to Benjamin Franklin, as early as the 18th century. Improving knowledge was something that some statesmen should consider. Benjamin Franklin's image appears on US 100-dollar bills (Figure 1.1).

Figure 1.1
Benjamin Franklin (1706–1790): American scientist and statesman

We have two options to choose from:

- Continue wasting money and time in injection molding plants by following the empirical method.

- Invest in improving the knowledge of people who define, control, and participate in the processes, by holding training sessions and seminars and by applying scientific injection molding methodology.

1.2 Scientific Injection Molding or Injection with Advanced Methods – What Is It?

The term and some of the methods that make up scientific molding were developed in the USA, as early as the end of the 20th century. The goal was to develop concepts that would contribute to a better understanding of the injection molding process for thermoplastics through the application of scientific methodology, and to establish robust, repeatable processes with little scrap and optimized cycle time and costs.

This methodology is supported by tests and analyses carried out on the machine that allow decisions to be made "on the front line" to define the processes. It is based on the scientific method.

It is also supported by various types of tests and tools that allow the user to analyze the injection process on the same machine and define robust process parameters. Additionally, it makes it possible to define the process inputs and outputs and repeat the optimal process in different production batches and even in different machines or different plants. It therefore allows technicians in daily contact with processes to improve them and get the most out of the performance of machines, molds, and materials.

In the USA, there are currently various training centers, consulting firms, and professional consultants that offer training in these methodologies for technical staff. Despite being recognized, this methodology is less widespread in Europe – with few technical centers providing training based on these scientific molding concepts (although in recent years, an outstanding effort has been made in this direction).

The scientific injection molding process is based on two main pillars:

- A **cultural change** from the empirical method to the scientific method
- Training, learning, and the application of the tools proposed by this methodology

1.2.1 The Cultural Change Involved Moving from the Empirical to the Scientific Method

Process preparation is elementary to the application of this methodology. It is about moving from the empirical method, where our decisions are based on experience, repetition, and trial and error (falling down and getting back up), and where, once a process configuration decision is made, we do not criticize it or wonder how we obtained the information, nor do we test hypotheses. Such decisions are based on perception (e.g. we simply know that if we touch fire we get burned, but we are not interested in knowing why).

In contrast, the scientific method is based on research, establishing a hypothesis, carrying out tests following a procedure, analyzing the data obtained to reach conclusions, and finally – something that is often not performed in injection molding plants – summarizing the information and disseminating it for possible follow-up.

The use of the empirical method (i. e. trial and error) means resources are used unproductively in companies in the mold-testing phases, later in the production phases, and also in derivative phases of the process. When robust processes are established, technicians' time spent dealing with process deviations, scrap generated during production and possible quality issues or complaints that may occur is reduced drastically, justifying the cultural change that must be made in injection molding plants.

1.2.2 Training and Applying Scientific Molding Tools

It is essential to train people who are in daily contact with the process and must define the conditions of the process. They should be trained in training centers by teachers who are experts in this methodology and, most importantly, its implementation. The training should be provided for all "key" team members.

Those of us who know (directly or indirectly) the plastic injection process have been able to observe the high number of variables involved. This makes injection molding a difficult process to understand, with multiple interrelated factors and high complexity, both in defining and establishing processes and in solving problems that may arise during manufacturing.

There are multiple root causes for problems or deviations that may occur during manufacturing or product development and they need to be controlled. The root causes may be classified into four major categories:

- **Material:** Material variability (right material selection, contamination, pre-drying, regrind level, etc.)

- **Design:** Part and mold design issues, related to the part (thickness changes, draft angles, etc.) or related to the mold (gate position and size, polishing, venting, cooling, etc.)

- **Process:** Process variability and issues (injection parameters, human error, packaging, etc.)

- **Machine:** Machine variability (machine inprecision machine performance, mechanical elements, peripherals, etc.)

These four major variability and issue groups are like the four wheels of a car: if any one of them is not properly inflated, aligned and in good condition, the car will not run properly and so the car will not perform optimally. All four must be in perfect condition and well-aligned to be successful. In injection molding, these four variability categories should be correct, aligned, and controlled.

In injection molding plants, we usually use the third "wheel" (process) to fix variability or deviation issues which originate from one of the other three (material, design, machine). How many times have we seen processes pushed to the limit (mold temperature, injection speeds, etc.) to address a problem that originates from another "wheel," such as the design of the part or mold, gate dimensions, inadequate material selection, or a machine that is not up to the task? It is often the "process" that is the wildcard; we tend to try to fix problems caused by another problem group (material, design, machine) rather than fixing the problem and solving it at its root cause.

1.2.3 Some Definitions

- Injection molding technician and troubleshooter

  Follows actions and procedures learned from experience

  - Changes several parameters at once, without waiting for the results of previous changes
  - Rarely documents performance changes, ratings, historical data, results, etc.
  - Lack of method can cause damage to the technical equipment, molds, etc.
  - Never uses specific data

- Science

  Science is the systematic pursuit of knowledge obtained through the scientific method to obtain general conclusions. According to the Dictionary of the Spanish Language (*Real Academia Española*), it is the set of knowledge obtained through observation and reasoning, systematically structured and from which general principles and laws with predictive and experimentally verifiable capacity are deduced.

- Intelligence

  Intelligence is the ability to learn and apply facts and skills, especially when this ability is highly developed. Intelligence is the ability to solve problems.

The role of an advanced technician is encapsulated by "intelligence", "science", and "technician and troubleshooter":

> Advanced technician – a trained technician using their knowledge and highly developed skills to define a robust process and resolve process deviations

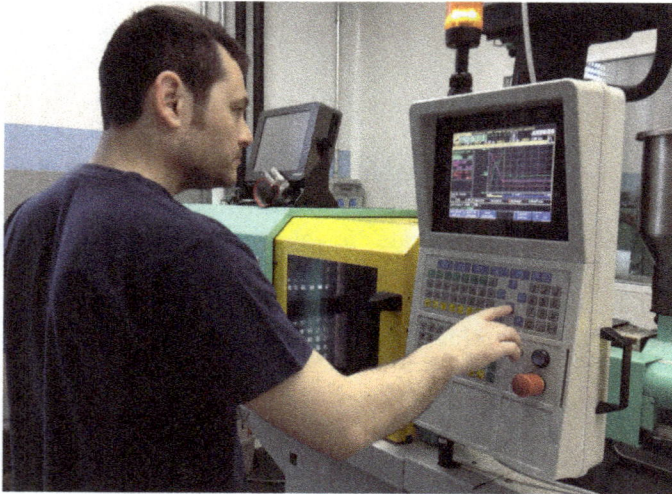

As mentioned in Section 1.2.1, with the application and understanding of scientific injection molding tools, there is a cultural change in the way we work, from the empirical method (trial and error) to the scientific method (applying knowledge in daily manufacturing and performing tests on the machine to obtain information, which helps us make decisions on the best parameterization of the injection process). This change makes it possible to optimize the process, to define a more robust and consistent process, and in many cases, to understand why the processes have deviated from standard procedure.

An advanced technician always:

- Learns about the history of the process, mold, machine, material, and technology at their disposal

- Analyzes and defines what has changed when a process deviation has occurred

- Acts with knowledge, and carries out tests and trials to gather information and define a robust process

- Checks the results and outputs of each process change

One of this methodology's key features is that the process is repeatable on successive process productions, serving as a reference when the process is correct and working optimally. This requires that the process parameters and conditions be checked and recorded. The more information that is recorded, the more likely it is that the process is repeatable for future production processes. This is where you must take process outputs into account.

1.2.4 Process Inputs and Outputs

The typical injection molding technician usually documents machine-dependent input values called "machine settings". When I ask in injection molding plants for the parameter sheets, I often find documentation that meticulously records machine settings; however, there is almost no output recorded.

The following are some examples of inputs:

- Various configured temperatures
- Configured injection rate
- Various configured times
- Various configured pressures
- Configured different strokes
- Configured clamping force

Figure 1.2 Input and output diagram

Advanced technicians monitor and record the machine's independent output data, called "process outputs" (Figure 1.2). Some process outputs to be monitored and recorded are:

- Temperature
  - Actual mold temperature
  - Actual cavity temperature
  - Actual coolant temperature, inlet and outlet
  - Actual feed-throat temperature
  - Actual air dryer temperature

- Time
    - Actual mold-filling time (the most critical time of the process)
    - Actual holding pressure time
    - Actual metering time
    - Actual gate sealing time
    - Actual cycle time
- Pressure (Figure 1.3)
    - Filling pressure at V/P switchover
    - Actual specific holding pressure
    - Actual specific back pressure
    - Delta P pressure
    - Machine intensification ratio
    - Cooling system pressure and flow
- Weight (Figure 1.4)
    - Cavity weight
    - Cavity weight at V/P switchover
    - Final cavity weight after gate seal
    - Total shot weight (runner included)
- Additional data
    - Cushion
    - Dimensions per cavity
    - Mold filling balance
    - Actual clamping force
    - Photographs of defects
    - Various observations, comments
    - Other

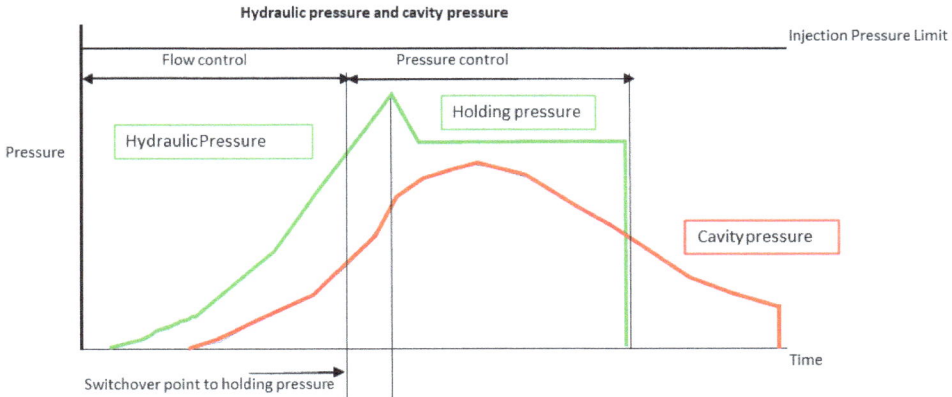

Figure 1.3 Hydraulic and cavity injection pressure graph

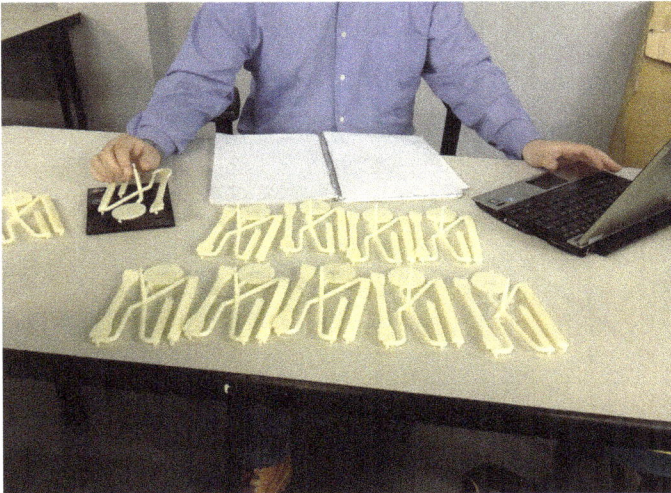

Figure 1.4
Systematic weight
analysis

If the process is well-documented, when process deviations occur, the data recorded can help the advanced technician to identify what has changed in the process. Furthermore, the advanced technician:

- Knows the background of the process, mold, material, peripherals, etc.

- Knows the properties and characteristics of the processed material

- Identifies the physical causes of the process changes or part defects through the process outputs

- Acts with knowledge and intelligence

- Checks the results and changes in the outputs every time that one parameter is changed

An advanced injection molding technician considers and calculates outputs, such as:

- Residence time

- Peripheral screw speed

- Injection unit intensification ratio (Figure 1.5)

- L/D ratio and injection unit utilization (%)

- Delta P

- Coolant flow rate (Reynolds number)

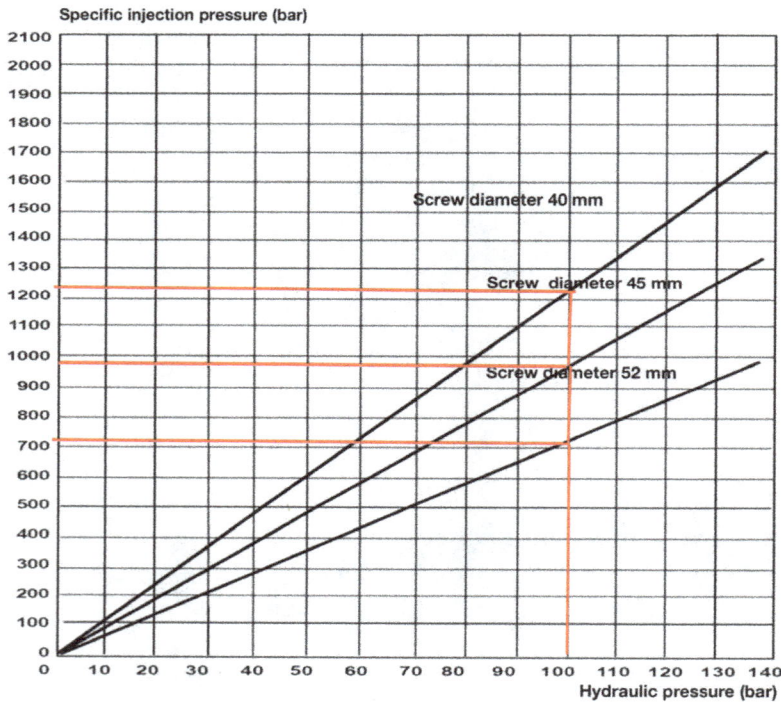

Figure 1.5 Graph showing intensification ratio for different screw diameters

While a typical injection molding technician wonders "Which parameter setting should I change to address the defect or process issue?" and "Can I find solutions in universal troubleshooting guides?", an advanced injection molding technician wonders "What changes have happened in the process and the outputs to cause the problem?" and "What is or what may be the physical root cause of the defect or process issue?"

Documenting the entire process and its changes is crucial to this type of methodology and process analysis.

1.2.5 New Tools for Process Definition

An advanced injection molding technician can evaluate the process, if necessary, through different tools, such as:

- In-mold rheology or relative viscosity test (Figure 1.6)
- Gate sealing time analysis (Figure 1.7)
- Process window determination (Figure 1.8)
- Injection speed linearization or injection speed behavior – analysis of compliance (Figure 1.9)
- Pressure loss analysis (Figure 1.10)
- Delta P analysis (Figure 1.11)
- Gate shear analysis (Figure 1.12)

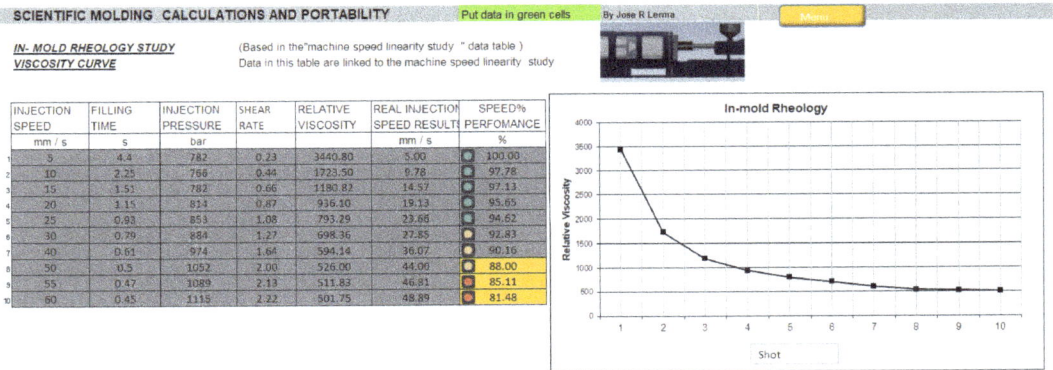

Figure 1.6 In-mold rheology method, data, and graph

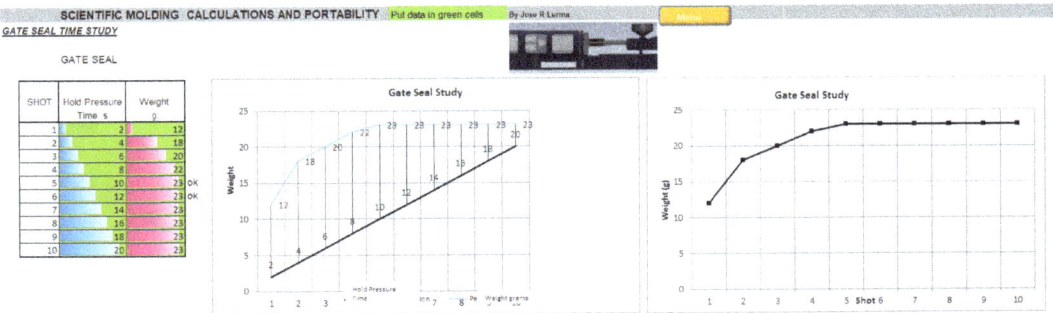

Figure 1.7 Graphs for analyzing the gate sealing time

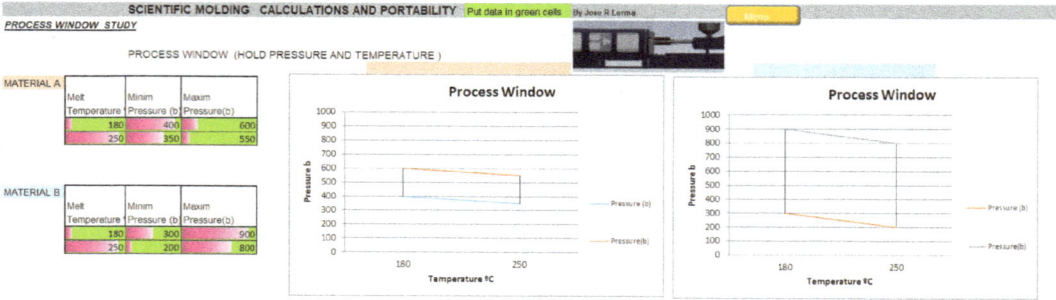

Figure 1.8 Graphs for determining the process window

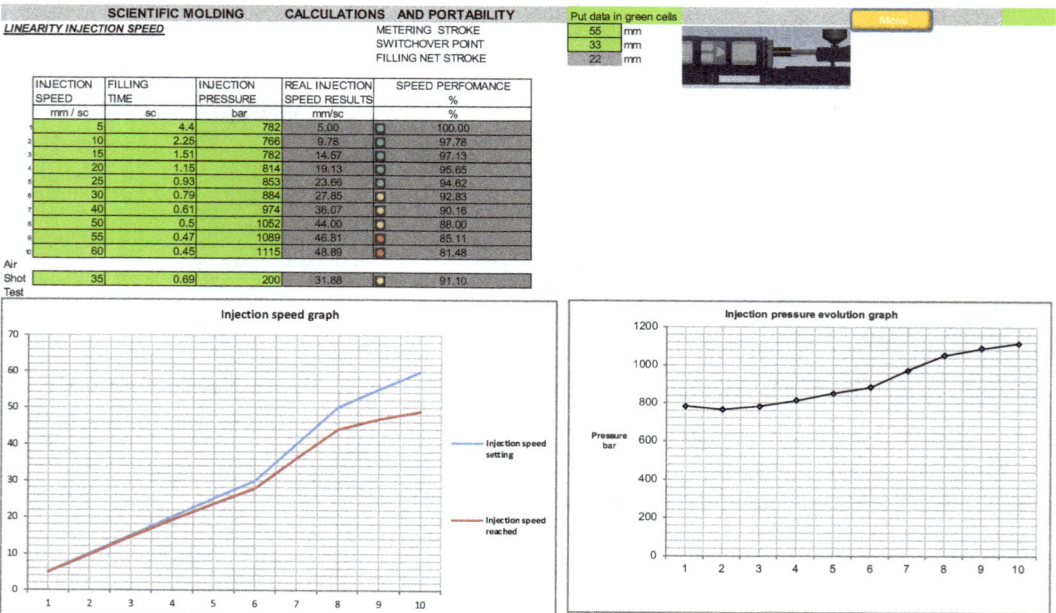

Figure 1.9 Graphs for linearizing the injection speed

Drop Injection Pressure Distribution

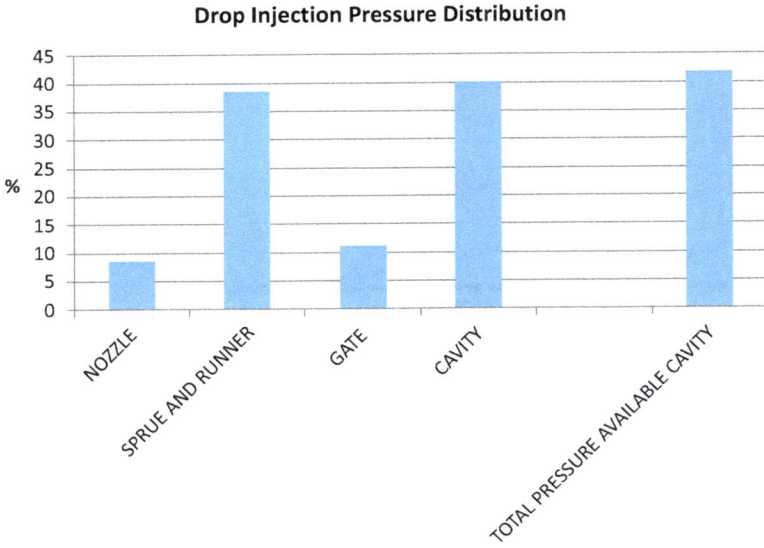

Figure 1.10 Graph for analyzing pressure loss

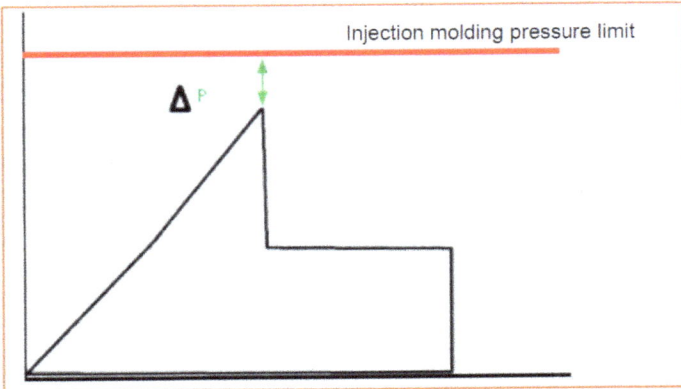

Figure 1.11 Graph for analyzing Delta P

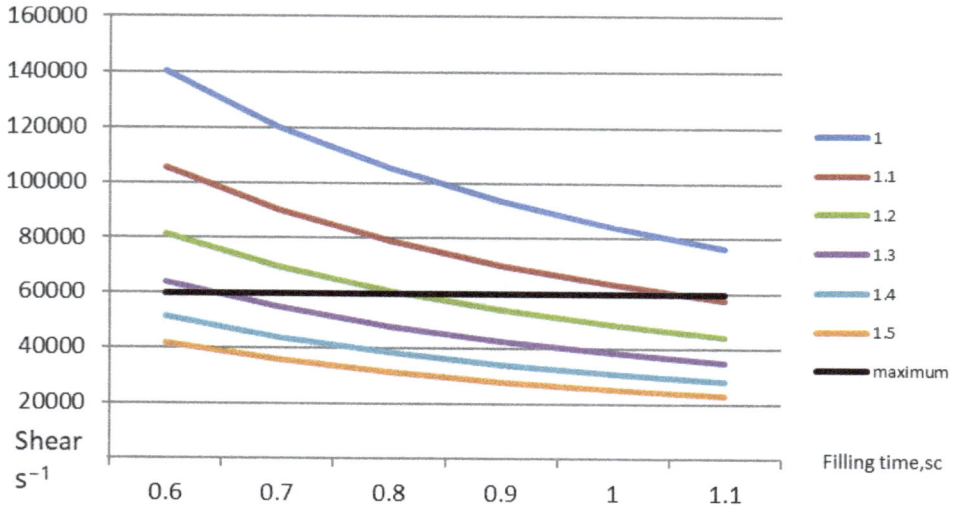

Figure 1.12 Graph for analyzing gate shear with different injection filling time and different gate diameter

Finally, an advanced technician can transfer the parameters defined in a robust and productive process to other machines with different characteristics by using and applying "portability" through the "Universal Parameter Sheet" (Figure 1.13).

These are only a few examples of the tools included in this methodology that can usually be used during the initial setup of a molding process or when deviation and issues arise. The time invested in carrying out these tests and taking advantage of the information they provide is a good investment for better productivity, less scrap, and a more robust, consistent, repetitive injection molding process.

These tools and much more are explained in detail on the following pages, each with a supporting spreadsheet.

SCIENTIFIC MOLDING		CALCULATIONS AND PORTABILITY			Put data in green cells	Menu
PART NAME		MATERIAL	ABS		DATE	
MOLD NUMBER		MELT DENSITY	0.95 g/cm3			

MACHINE A				MACHINE B		
MACHINE A	150	Tn		MACHINE B	480	Tn
SCREW DIAMETER	50	mm		SCREW DIAMETER	70	mm
METERING STROKE	115	mm		METERING STROKE	54.70	mm
DOSAGE VOLUME	225.8	cm³		DOSAGE VOLUME	225.80	cm³
WEIGHT	200	g		WEIGHT	200	g
MAXIMUM METERING STROKE	300	mm		MAXIMUM METERING STROKE	500	mm
RESIDENCE TIME	2.61	minutes		RESIDENCE TIME	7.68	minutes
METERING : SCREW DIAM RATIO	2.30	times		METERING : SCREW DIAM RATIO	0.78	times
INJECTION UNIT UTILIZATION(%)	38.3	%		INJECTION UNIT UTILIZATION(%)	10.9	%
CYCLE TIME	45	sc		CYCLE TIME	45	sc
FILLING TIME	2	sc		FILLING TIME	2	sc
REAL INJECTION LINEAR SPEED	55	mm/sc		REAL INJECTION LINEAR SPEED	28.06	mm/sc
REAL INJECTION VOLUME SPEED	107.99	cm³/sc		REAL INJECTION VOLUME SPEED	107.99	cm³/sc
INTENSIFICATION RATIO	10	:1		INTENSIFICATION RATIO	8	:1
SPECIFIC INJECTION PRESSURE	1400	bar		SPECIFIC INJECTION PRESSURE	1400	bar
HYDRAULIC INJECTION PRESSURE	140	bar		HYDRAULIC INJECTION PRESSURE	175	bar
BACK PRESSURE	10	bar		BACK PRESSURE	12.5	bar
MOLD FILLING STROKE	110	mm		MOLD FILLING STROKE	56.12	mm
MOLD FILLING VOLUME	216.0	cm3		MOLD FILLING VOLUME	216.0	cm3
SCREW RPM	80	rpm		SCREW RPM	57.14	rpm
SCREW PERIPHERAL SPEED	0.21	m/sc		SCREW PERIPHERAL SPEED	0.21	m/sc
COOLING TIME	15	sc		COOLING TIME	15	sc

Figure 1.13 Calculation of process portability to different machines

1.2.6 Scientific Injection Molding or Injection with Advanced Methods

Scientific injection molding definition:

> The application of the state of structured knowledge through reasoning and experimentation in the field of plastic injection molding

Or, in other words:

> The application of science and intelligence to plastic injection molding

All the scientific injection molding tools (Figure 1.14) have been conceived to dominate and control the injection molding process with the goal of defining repeatable, robust, and consistent processes (consequently, productivity will improve).

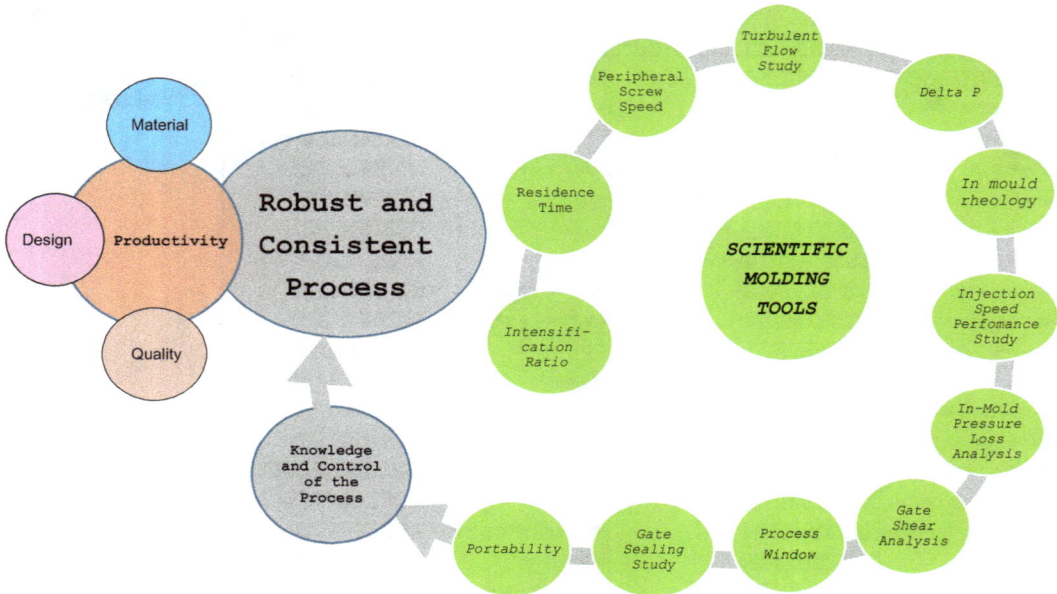

Figure 1.14 Scientific injection molding tools – improving productivity through knowledge of the process

In summary:

- A typical injection molding technician:
    - Focuses on machine parameters, machine inputs, and the empirical method
- An advanced injection molding technician:
    - Focuses on injection molding from a material perspective – output is the key; this is called the "black-box approach"
    - Focuses on machine inputs and outputs
    - Records data, process history, molds, materials, peripherals
    - Optimizes and acts from a material perspective, treats the cause, not the symptom

This is the major paradigm for many injection molding technicians.

Injection molding technicians will improve their skills through trainings, seminars, and doing series of experiments that add to their experience and skills developed with knowledge. By analyzing the data and results of the experiments, advanced technicians can identify the weaknesses of a process to define a consistent, robust process. These advanced injection molding technicians should focus on finding and discovering the root cause of the problems and weaknesses of the process.

1.3 Who Should Be Trained in This Methodology and Where?

For the successful implementation of scientific injection molding in injection molding plants, the "key team" must be trained in the fundamentals of this methodology.

There are many cases of failure in the application and implementation of the scientific injection molding system caused by the improper selection of the personnel involved. A common situation is when a technician attends a scientific molding seminar and becomes enthusiastic about the tools and methods proposed and explained in the seminar. The technician returns to the injection molding plant highly motivated to implement the tests and methodology with new molds, process deviations, process issues, etc., and practice the methodology learned. Since the rest of the organization knows neither the tests nor the methodology, and does not correctly understand the results, this trained technician is effectively alone in implementing the new system, and the inevitable result is failure.

In all cases of successful implementation of the methodology, the training is extended to all "key team" members, so that everyone in the organization can speak the same language and understands why the actions are taken, understanding every one of the tools or methods that the system proposes. After this "key team" training, the internal technical team discusses – as a team with a better level of knowledge – the approval of new projects, the validation of new molds, the improvement of existing ones, the correction and tuning of process deviations, and the analysis of the root causes.

If budget cuts are likely, cutting back on personnel training should be a last resort. The human team of an injection molding plant is like a chain – it will only be as strong as its weakest link. Well-oriented and carefully selected training acts as a reinforcement for this chain when it is applied in the plants on a daily basis.

The names of the jobs or roles within injection molding plant organizations may vary with each plant, sector, or even country. However, the role or task within the organization is common to all of them. These are the roles that must be trained in scientific injection molding:

- **Injection molding technicians** – technicians who carry out the start-up of production by following the established parameter sheets and who, where necessary, modify and adjust some parameters and tune the process for proper quality and productivity during injection molding production series

  Once trained, they will be able to understand machine settings and outputs and therefore record and focus on what is really important in mastering and controlling the process. They will be the users or internal customers of the process parameters established by process engineers. They must fine-tune these parameters, adapting them to the changing situations of the process during different production campaigns, but always using the scientific method, not the empirical method (trial and error).

Without a well-trained injection molding technician, it is not possible to successfully implement the scientific method because, even if we initially establish a stable, robust process, successive productions, batch changes, different machines etc. will quickly result in an uncontrolled process that is completely different from the initial methodology. This will eventually lead to the non-application of the methodology learned, due to the pressure of everyday work and a lack of confidence in the results.

In addition to being trained in scientific molding, these technicians must undergo a cultural change from molding from a machine perspective to molding from a material perspective (i.e. abandoning the empirical method based on trial and error and switching to the scientific method).

- **Injection molding process engineers** – technicians whose job is to define the injection processes from the beginning of the project, from the first trials

The most important job, from the process perspective, is to be responsible for its definition from the start. This technician should be in contact with the product designer, the mold designer, the mold maker and often the manufacturer of the raw material to anticipate the potential problems that will undoubtedly arise if this worker does not participate in the previous design steps. This role is key because it defines process conditions throughout the entire project. These technicians must define robust process conditions and a process window with a large molding area; otherwise, the project will be unproductive and possibly unprofitable for a long time.

It is therefore essential to train these technicians in scientific injection molding. In this way, the recommendations in the previous steps will help to achieve a robust, productive, and consistent process. The use of scientific injection molding tools will help these technicians understand and master the process. It should not be forgotten that these technicians, well-trained in scientific molding, can accurately determine the ideal machine for a certain product or mold using scientific injection molding tools.

These technicians must also undergo a cultural change from using the empirical method, and instead opt for the methods proposed in scientific molding methodology.

- **Quality engineers** – technicians working in the quality control department

Their training will have a direct impact on understanding the process, understanding the actions that result from mold testing, and even suggesting some process improvements to be made. For these technicians, understanding the process, the root causes of defects, the causes of process deviations, knowing if a process is focused, etc. is a vital element of their responsibility. They can provide – if they know the process and are trained in scientific molding – improvements in the initial stages of early production testing.

By understanding the tests that are carried out with their collaboration during mold testing or by analyzing the process engineers' reports, and if these technicians know the scientific molding methods and tests, they will be able to understand the weakness or robustness of the established process. This understanding leads to proper definitions of quality control plans for each process, application of more thorough controls where required, and inspires confidence in the process where robustness has been demonstrated.

These technicians trained in scientific molding can collaborate in and/or lead process improvement or deviation analysis teams from the point of view of understanding the outputs and real changes that have occurred as a result of a process deviation that leads to the production of defective parts.

- **Simulation engineers** – technicians who carry out computer injection molding simulations with one of the current simulation systems in the initial stages of a project

These technicians often find that the simulation results are not reproducible in the machine – in the real process, with the mold and material in the injection molding plant. It is important for these technicians to be trained in this methodology, as it will help them understand these differences and be able to correct them with the aim of developing skills to enable the simulation carried out to resemble the real process more closely. Training simulation engineers in this methodology could be the missing link that these technicians need in order to connect simulation theory with the reality of the machine's injection molding process.

If simulation engineers optimize the process, mold, inputs, gates, cooling, etc. from the beginning during the simulation stage, time and resources will be saved and manufacturing will be much more successful. This is the main reason why it is very important to train these technicians in this methodology.

- **Molders, responsible for mold maintenance**

This job is sometimes outsourced and sometimes integrated into production plants, in line with each plant's strategy.

It is vital that these technicians understand and become familiar with scientific mold testing. Good training will help them understand the possible mold weaknesses that can be identified with this methodology or understand and agree with some mold modifications proposed by the process engineers through the scientific molding method. If they do not know the tests, methodology, and fundamentals, they will not believe and will not understand the results obtained by the process engineers.

These technicians will define the preventive maintenance of the molds, the maintenance schedule, the molds' expected lifespan, and the spare parts to have available, for the purpose of controlling and reducing future interruptions arising from breakage, breakdowns, wear and tear, etc. With good training, these technicians will work as a team and be "on the same page" as the process engineers.

If the molds' maintenance team is not trained and does not understand these changes, it is difficult to convince them. In my opinion, these technicians are a fundamental part of the application of scientific molding in injection molding plants both in the testing and industrialization stages of new molds and in the analysis of process deviations and actions that can be implemented.

- **Product designers**

This role is the one that turns an idea or a concept into a particular product. Product designers should rely on the material engineer from the material manufacturer or supplier for proper material selection. An error made here can become a big problem throughout a product's lifespan. Product designers' focus is usually on the product's functionality and ensuring that it meets the required function and design, but they should not neglect the machine, the mold, or the productivity of the process.

If product designers are trained in the scientific molding methodology, they will be much more aware of the process during the design phase. Their focus will also be on achieving the largest possible window or molding area in the process. Actions such as draft angles to eliminate thickness accumulation zones, thickness changes, sharp corners, etc. are critical to a robust process, and so product designers must keep them in mind. Scientific molding tests such as the process window or pressure loss analysis during the cavity filling will help these engineers tweak the design as necessary to further focus the process into the largest process window possible.

- **Mold design engineers**

Mold design engineers must know what elements are necessary in the mold design stage to help the injection molding technician define a robust process with a large molding process window. They must liaise with the material supplier to understand the material's behavior (e.g. shrinkage expectations, best shapes and dimension recommended for channels and gates, venting, mold steel coatings, recommended mold steel alloys, cavity steel hardness, etc.).

It is recommended that the proposed mold design be validated or reviewed (before starting to machine steel) by other personnel, such as the product design manager or the process engineer, for potential improvements to the mold design at the initial stage of mold definition. Special or specific characteristics of the machine(s) that will inject the parts should also be taken into account during the mold design stage in order to optimize how the mold "fits" into the machine. Scientific injection molding tests and studies, such as gate shear analysis or coolant flow rate, are very useful for helping technicians to improve mold designs.

- **The material supplier**

  The material supplier knows the most about the material to be used and must therefore be involved in new developments. An excellent knowledge of the scientific molding process and methodology will help them to provide or propose solutions from a material perspective (e. g. alternative materials, different rheological behavior, materials with additives to improve processability, nucleated alternative materials, etc.). For example, an understanding of the rheological behavior of a set of material–mold–machine and process conditions is important for understanding the behavior of a given material and possible alternatives.

- **The sales and logistics team**

  An injection molding plant sales team trained in these scientific molding methodologies will consider the capabilities of their injection machines when choosing whether or not to accept a new project. The following are relevant concepts to be considered:

  - Residence time calculation
  - Injection unit utilization (%)
  - L/D ratio
  - Theoretical clamping force
  - Specific injection pressure, material viscosity, thickness, and flow length

  All these calculations and factors concerning the selection of the right injection machine will be considered (which are directly related to the cost of the part) by a properly trained sales team to avoid wrong machine selection, costing errors, and a difference between the expected cost and the actual final cost during the production of parts.

- **Engineering and design**

  Those involved in engineering and design are generally crucial. Anticipating production issues makes it easier to eliminate problems during this early stage rather than in more advanced stages. Molds, assembly tools, plant layout, machinery, auxiliary equipment, peripherals and so on can be your allies or your worst enemies if you do not adequately prepare them.

  Once trained, engineering and design personnel will be able to see the full picture of the problem and will have tools to analyze causes and solutions, if necessary. These technicians will participate in the validation of molds, tools, and processes, following the scientific method and knowing where the limitations are (what the process window is, etc.), therefore avoiding wasting of time and money.

- **Others**

  The application of scientific injection molding affects virtually all injection molding plant departments, and all of them must be trained on the proper implementation of the system. This methodology will make the company more productive and profitable only if all plant staff get on board. The professional activities in the injection molding plant will also be made easier, leading to successful implementation.

The entire organization must be trained in scientific molding and understand the vital importance of defining and having a large process window. If the process engineer sends a mold back to the molder after the first testing to make corrections, to tweak, and fine-tune a certain mold's design or appearance issues that prevent a suitable or expected process window, the molder team, along with its manager (mold engineers, mold maintenance), will fully understand the reasons and agree to rectify any issues.

It is not a matter of manufacturing 100 good parts, but defining a process that is capable of manufacturing millions of good parts. This is only possible if all the relevant departments are well trained in this methodology.

Previous Training... Where?

The necessary training for the proper implementation of the system can be carried out in different technical centers. It must be carried out by instructors with hands-on experience in these methodologies in the industry. From personal experience, I have seen that the best instructors are those who have worked in injection molding plants and have the experience and the skill of the "real" industry – be wary of theorists who have not set foot in plants.

Some leading centers for training or seminars are listed below. There are a number of options on offer (face-to-face or online and practical or theoretical).

- In Spain:
  - AIMPLAS, a leading technical center in Valencia (*www.aimplas.es*)
    - Seminars with theoretical and practical learning
  - ANDALTEC, a leading technical center in Andalucia (*www.andaltec.org*)
    - Seminars with theoretical and practical learning
  - Eurecat, a leading technical center in Catalonia (*www.eurecat.org/es*)
    - Seminars with theoretical and practical learning
  - AITIIP, a leading technical center in Aragon (*www.aitiip.com*)
    - Seminars with theoretical and practical learning
  - ASIMM website (*www.asimm.es*)
    - Scientific injection molding website with downloadable courses with different modulus
- Elsewhere in Europe:
  - RJG (*www.rjginc.com*)
    (Germany, France, UK)

- In the USA:
  - FIMMTECH (*www.fimmtech.com*)
  - JOHN BOZZELLI SEMINARS (*www.scientificmolding.com*)
  - ROUTSIS RTAINING (*www.traininteractive.com*)
  - PAULSON PLASTICS ACADEMY (*www.paulsonplasticsacademy.com*)
  - RJG (*www.rjginc.com*)
  - ORBITAL (*www.orbitalplastics.com*)
- In Mexico
  - RHL PLASTICS (*www.rhlplastics.com*)

Beyond training available in technical centers (either face-to-face or online), it is highly advisable to also make use of other resources (books, spreadsheets, etc.) that help you to understand the methodology of scientific injection molding. Some of them are:

- *Plastics Injection Molding*, Hanser Publications – Jose R. Lerma, *www.hanserpublications.com*

- *Robust Process Development and Scientific Molding*, Hanser Publications – Suhas Kulkarni, *www.fimmtech.com*

- *A Practical Approach to Scientific Molding,* Hanser Publications – Gary F. Schiller, *www.hanserpublications.com*

- *Runner and Gating Design Handbook*, Hanser Publications – John P. Beaumont

1.4 Where to Apply Scientific Injection Molding

This methodology can be employed in all plastic injection molding plants without exception. The more technical the sector and the more technical or complex the materials and applications, the more sense it makes to apply the methodologies explained here. However, no matter how "commodity" our materials, products, or applications are, it is still interesting to strengthen and improve the productivity and efficiency of our plastic injection processes.

It is important to apply these studies to new molds or projects, so that they get up and running correctly and their processes are defined in a robust and productive manner and the parameters have been validated through tests and studies explained in this book and not by the experience of the technician process.

It is also very interesting to apply this methodology to molds already in production, to analyze the process and undoubtedly improve it. We can then compare the results of:

- Time-cycle improvement

- Scrap or rejection ratio reduction

- Process repeatability and consistency

- Reduced impact of the external factors of the process

- More production stability in different machines with different batches of material

- Better analysis of process deviations that may occur

- Plant team on the same page and focused on a methodical and scientific way of defining and controlling processes

Developing and defining a plastic injection molding process is not simply about putting plastic in a mold. Instead, it is about ensuring the following:

- The mold is placed in the right machine, based on machine suitability calculations for the intended mold.

- The machine is in perfect working condition, as verified by on-site testing.

- Cooling systems have identified each connection (each tube or bridge) in respect of flow rates, Reynolds number, coolant temperature, and everything that enables repeatability of the cooling system connections in subsequent series.

- Carrying out experiments and tests deemed necessary to define a basic and robust process, detecting the weaknesses of the process and eliminating them.

Doing all this will ensure that injection molding plants will be more competitive in an increasingly complex and demanding environment.

Without the application of the scientific method (applying the empirical method instead), injection molding plants are often – from a process perspective – a mixture of a lack of method and a lack of training or knowledge. In many injection molding plants, process technicians are immersed in a kind of "battle" as they constantly deal with problems, due to constant process variations. In most cases, either because of a lack of time or knowledge and training, these are temporary solutions without either analysis or conclusions of the actions that attack the root cause of deviations, or studies that help analyze deviations and ultimately define a robust and consistent process. In these injection molding plants, it is common to hear the process technicians say, "All the parameters are the same; it's the same as always, yet the parts are not right", "We have tried everything and nothing works", or "You won't be able to do any better". In these plants, it is common to see process engineers or process managers go from machine to machine, tweaking the process conditions, constantly dealing with problems. Plants with variable processes exist because the processes are fixed in the absence of analysis and based on the empirical method. In these plants, processes are constantly corrected without first being analyzed and the rapid process tweaks become the usual long-term remedies.

The plastic injection molding industry and competition means that customers increasingly require more quality, which requires the continuous improvements to processes, more stable and repetitive production, and, of course, competitive efficiency in terms of cost and scrap. This is where the concepts that appear in this book come in – you cannot ignore having a perfect knowledge of your machines, your molds, your plastics, and, above all, the correct process definition. This is part of the indispensable continuous improvement that must be promoted in injection molding plants.

1.5 Failures in the Implementation of Scientific Injection Molding

I have been training injection molding technicians since 2005 and, since 2013, I have been training technicians in the methodology of scientific injection molding or scientific injection (also called injection molding with advanced methods). Once these technicians learn the methods and tests that this methodology proposes, they are responsible for applying them in the real industry – during the demanding everyday work of the plants. Once the technicians finish their training, they are highly motivated and eager to apply what they have learned and check that everything they have learned will help them in their daily work.

When trying to apply these methods, they often encounter situations and problems that cause them to return to the old empirical method that was used prior to the training, therefore eliminating the possibility of improvement and application of the improvements learned. This often leaves these technicians somewhat frustrated.

Some of the causes of failure in the systematic implementation and application of the methods that scientific molding proposes are as follows:

1. Many of the failures in the implementation of this methodology are due to a lack of knowledge of the real precision level of the injection machines in which we are applying them. No matter how good the methods, tests, training, and strategies which we want to use, if the machine does not have a good performance, good calibrations, no wear and tear, etc., there will be no reliability and we will not get good results. We must therefore know the real precision level of our machines, how they react, how they respond, whether they are calibrated or if they are not running under perfect conditions to manufacture the right parts productively and efficiently. We cannot achieve improvements with an uncalibrated, worn-out, unreliable machine.

2. As explained in Section 1.2.2, we can establish four major groups of variability and issues that could arise in the injection molding process, which are like the four

wheels of a car – if any of them is not in good condition, the car will not perform well. We need to have all four variability and issue categories in perfect condition in order to succeed: **material**, **design**, **process**, and **machine**. If the three wheels that make up the process are not correct, the process alone will not be robust. At best, you will be able to manufacture parts but with little chance of stability and productivity. For example, producing parts in an inadequately-sized machine with an injection unit too large for the molding part produced will inevitably cause a long residence time, polymer degradation and material property loss, or even a thin-walled part that requires a higher injection pressure than that available in the machine injection unit or a worn injection unit, etc. There are many situations outside of the process where an injection molding technician can define and achieve a robust process by applying any methodology; it is useless trying to apply advanced or scientific molding techniques in certain situations.

3. It is not possible to apply some methodologies to certain molds, but this should not affect our willingness to apply these methods where possible. For example, it is not appropriate to carry out a relative viscosity test or in-mold rheology study on molds with inserts, whether metallic or plastic, as these inserts can move when the test is performed at high injection speeds. Nor is it appropriate to carry out these tests on molds with very thin walls that require maximum pressure from the beginning of the cavity filling or in parts with very little screw movement for the metering and injection stroke, nor in parts that create demolding problems if they are not completely filled.

4. It is not possible to analyze the gate sealing time in such cases as:

    a) Very soft materials and TPEs, because we can make material enter the cavity for a long time, even if this material has become solid or semi-solid during the holding pressure stage, generating tension and exfoliation near the gate.

    b) Hot runners, because they do not have a cold gate with the possibility of sealing by cooling. See Section 7.2.8 for further details of how to proceed in this case.

5. Starting to define processes in new molds is an even greater challenge when there is no in-depth knowledge of the methodology of scientific molding, so is recommended starting with molds that we are already familiar with, where we can assess whether the adjustments made give us robustness and stability, less scrap, better cycles, etc. With familiar molds, we will be able to learn these techniques better and then apply them to new molds in new projects in a second phase. In addition, the results will encourage more widespread implementation within the plant and will convince even the most reluctant technicians.

6. Last but not least, a risk of failure is a lack of support from management in training and seminars, as they are the preliminary, fundamental step in the application of these scientific injection molding methods in injection molding plants. As with

any change in working methods, this must be heavily promoted and supported by management, who must be at the forefront of believing in and convincing others of the benefits that these methods bring to the companies that apply them. Without the support, patience, and drive of senior management, the application of scientific molding will not be possible. It is therefore important to involve management and convince them with data that this implementation should be supported.

2 Knowing Our Machines

In order to apply scientific injection molding and its methods, it is necessary to know our machines, which means that not only do we have to know the characteristics of our machines (some of these characteristics, which we will see later, are unknown in many injection molding plants), but we also have to know the performance of the machines regarding precision repeatability, calibration, linearization, etc. Injection molders think they know their injection machines well, but many of them are actually so busy in their daily work that they cannot see the wood for the trees.

If we want our injection molding machine to do exactly what we ask of it, we must know the machine perfectly (its precision level, control, characteristics, responses, adjustments, calibrations, misalignments (if these are controlled), etc.). Bear in mind that these characteristics, outputs, calibrations, and so on may differ from machine to machine (even if they are the same model, age, size, and manufacturer).

As mentioned in the previous chapter, many of the failures in the implementation of this methodology are due to a lack of knowledge of the machines to which we apply it.

2.1 The Injection Molding Machine

The injection molding machine can be divided into two main units: the clamping unit and the injection unit. The clamping unit is characterized by, amongst others, the clamping force, the moving platen stroke, the tiebar spacing, the minimum and maximum mold thicknesses, and the clamping speed. The injection unit has several characteristics, such as the screw diameter, maximum pressure, L/D ratio, compression ratio, plasticizing capacity, maximum injection volume, heating power, and maximum injection speed.

Figure 2.1 Injection molding machine (Source: Wittmann Battenfeld)

2.2 Clamping Unit

The main characteristics of the clamping unit are the following:

- Clamping force
- Moving platen stroke
- Tiebar spacing
- Maximum and minimum mold thickness
- Platen thickness
- Ejection stroke

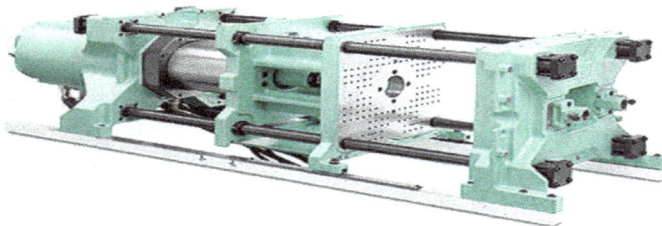

Figure 2.2 Hydraulic clamping system

2.2.1 Clamping Force

The function of the clamping force is to keep the mold closed so that it does not open due to the injection pressure thrust during cavity filling and packing. If the product obtained by multiplying the projected part area in the mold by the required injection pressure exceeds the clamping force, the mold will open and the tiebars will be over-stressed and elongate even more. They may exceed the elastic tensile limit of the steel and break or deflect the clamping plates.

There are various clamping systems (see Section 2.2.2):

- **Mechanical clamping systems** are stiffer than hydraulic clamping systems. Once the toggle system is locked, the maximum clamping force is applied and can only be released by the central hydraulic piston.

- A **hydraulic clamping system** applies pressure close to the center of the plate. Due to the compressibility of the hydraulic oil system, this system is not as rigid as a mechanical clamping system.

- An **electric clamping system** is a mechanical clamp activated by an electric motor instead of a hydraulic piston.

- The latest generation of mechanical toggle clamping systems also concentrates force close to the center of the moving platen.

Moving platen stroke:

A longer stroke makes the machine more versatile. In the hydraulic clamping systems, the total stroke of the piston is equal to the sum of the mold thickness and the maximum mold-opening stroke.

Tiebar spacing:

The spacing should be as wide as possible, provided that the maximum permissible bending stresses of the plates are not exceeded.

Fixed and moving platens:

These should be parallel. The weight of the moving platen and the mold must rest on the base of the bed and not on the tiebars.

Mold size relative to platen size:

As a rule of thumb, molds with a base area less than ¼ of the area of the moving platen should not be installed. **The projected area of the molds used should not be less than ¼ of the area marked out by the tiebars in the clamping plate.** If the mold area is smaller, the plates may bend more than recommended.

2.2.2 Clamping Unit Systems

According to their design, as mentioned in the previous section, the following sealing systems can be distinguished:

- Mechanical toggle clamping systems (see Figure 2.3 and Table 2.1)
- Hydraulic clamping systems (see Figure 2.4 and Table 2.2)
- Hydraulic two-stage piston systems (see Figure 2.5 and Table 2.2)
- Tiebar-less systems
- Electric systems

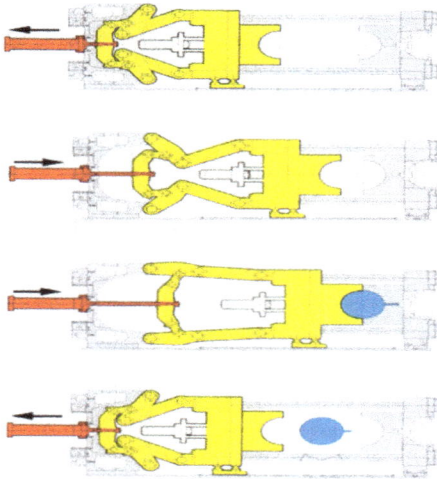

Figure 2.3
Mechanical toggle clamping system

Figure 2.4 Hydraulic clamping system (with central piston clamping force)

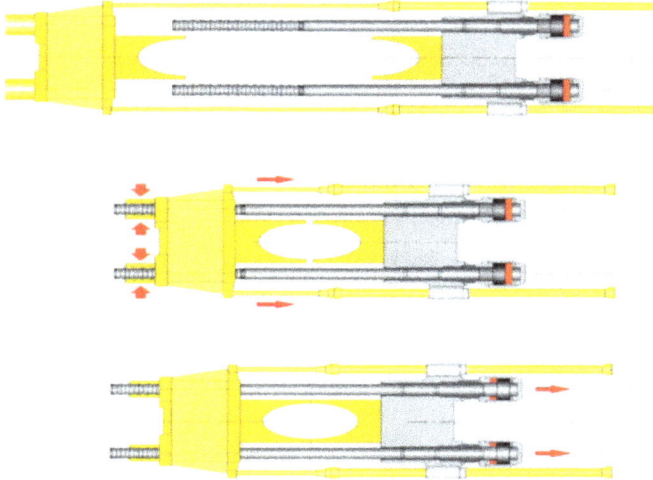

Figure 2.5 Hydraulic clamping system for large tonnage machine (with four lateral pistons)

There are several types of two-stage clamping systems that may be included in hydraulic clamping systems. The most common system is a mechanical lock in the machine made by two very small hydraulic cylinders that drive two locking parts which act on the central axis. Once the central axis is locked, high pressure is applied through a larger cylinder (pressure cylinder), which moves it only a few millimeters to provide the necessary preset clamping force.

Mechanical Toggle Clamping System versus Hydraulic Clamping System

Table 2.1 Mechanical Toggle Clamping System

Advantages	Disadvantages
Faster movements	More maintenance needed
Low oil flow	Short opening strokes
Low sensitivity to hydraulic leaks	Tendency to overload
Favorable kinematics	Less movement precision
Lower energy costs	More mechanical elements and more wear and tear

Table 2.2 Hydraulic Clamping System

Advantages	Disadvantages
Long opening strokes	Slower movements
Cleanroom application	Higher energetic cost
No breakage stress on tiebars	Unfavorable kinematics
Less mechanical maintenance needed	Sensitivity to hydraulic leaks

2.2.3 Theoretical Clamping Force Required

Materials will require different injection pressure levels depending on their viscosity.

Scale A: low-viscosity material (PA/PE/PP/PSB)

Scale B: medium-viscosity material (ABS/CA/POM/SBC)

Scale C: high-viscosity material (PC/PMMA/PPO/PVC)

In order to use the graph in Figure 2.6 as an aid in the calculation of the estimated clamping force, we need to know the flow path, the part thickness, and the viscosity of the injected material. In the example shown in red in the graph, for a flow path of 180 mm and a part thickness of 1.5 mm, we can see the specific pressure needed for the different materials on scales A, B, and C.

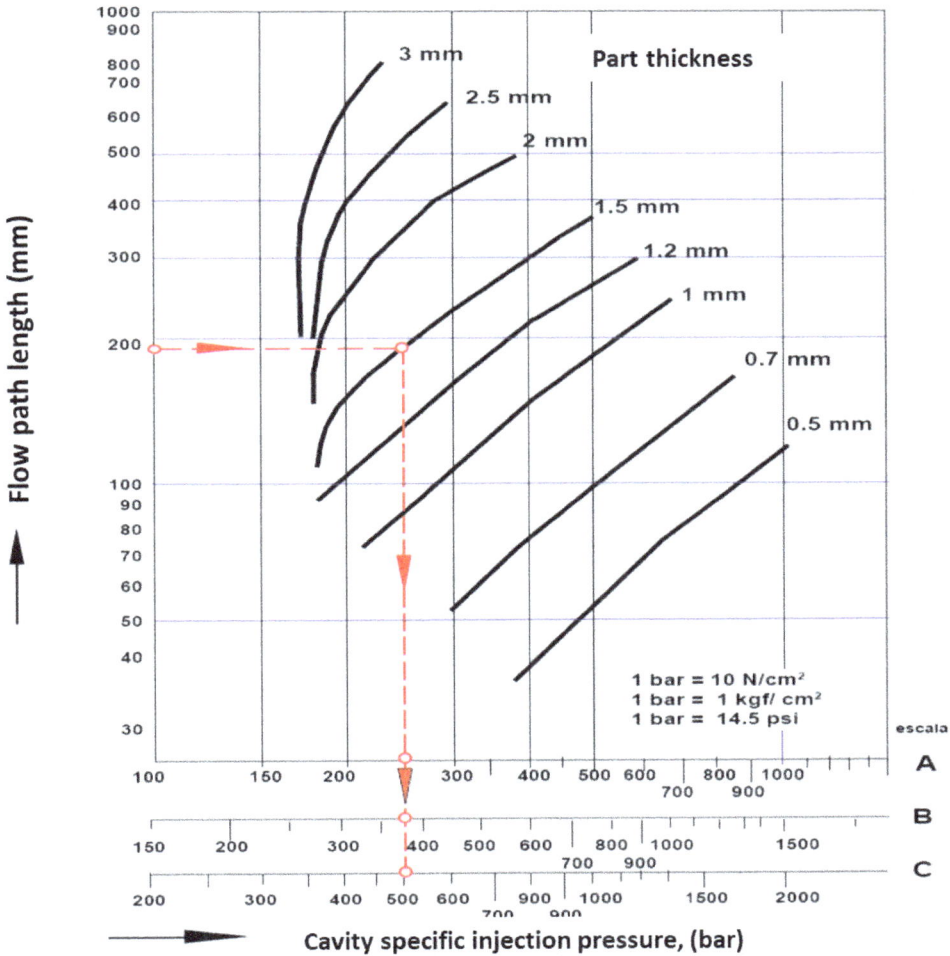

Figure 2.6 Graph of injection pressure as a function of flow path length, part thickness, and material viscosity

The pressure obtained from the graph must be multiplied by the projected area of the parts to be molded. We can then determine the total required clamping force by multiplying the projected area of the parts by the maximum pressure in the cavities. In the case of molds with slides, we should add 30% of the slides projected area to the projected cavity part area because the slides significantly increase the required clamping force.

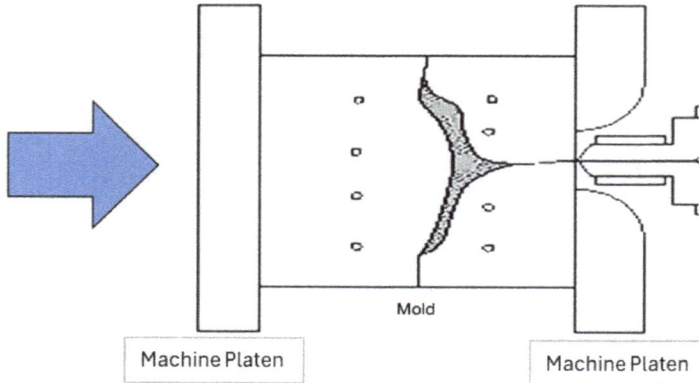

Figure 2.7 Clamping system diagram

The required clamping force equals the injection pressure (into the cavity) multiplied by the projected area:

$$\text{CLAMPING FORCE } [t] = \frac{\text{INJECTION PRESSURE} \cdot \text{AREA } [cm^2]}{1000}$$

2.3 Injection Unit

We require the injection unit to perform many **functions**, all of which are essential for a robust injection molding process. The following are the most important:

- Good plasticizing of the melt material, homogeneity of the melted plastic
- Metering volume repeatability
- Speed and pressure repeatability
- High movement and position precision

The injection unit (Figure 2.8) consists of the following main elements:

- Hopper
- Plasticizing barrel
- Screw
- Non-return valve or screw tip valve
- Screw tip
- Nozzle

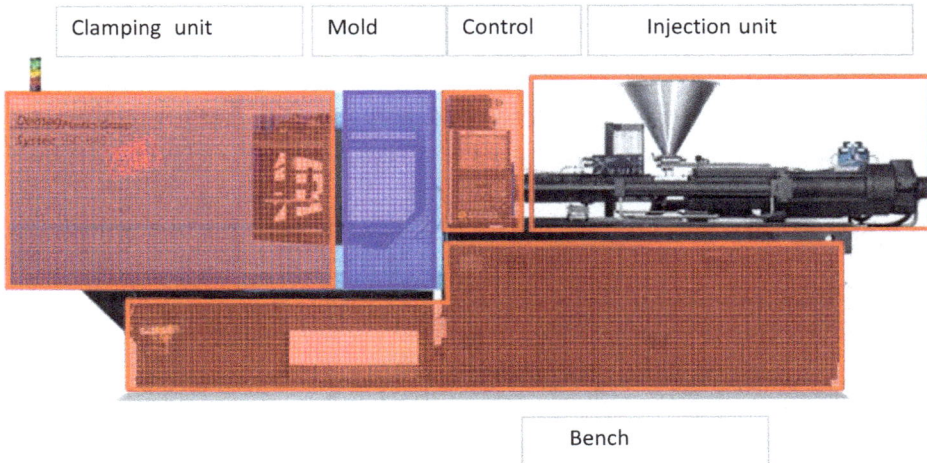

Figure 2.8 Injection unit and others parts of the injection machine

Injection Unit Characteristics

- L/D ratio

  The L/D ratio is the ratio of the length L to the diameter D of the screw.

  - Typical L/D ratio: 10 to 20 in injection molding processes
  - For ratios higher than 20: overly long material path, more homogeneity, but also longer residence time, possible degradation, etc.
  - For ratios lower than 20: possible lack of homogeneity in the melt

- K ratio or compression ratio

  The K ratio is the ratio of the volume of a barrel in the screw in the feed zone near the hopper (h2) to the volume of another barrel in the metering zone near the valve ring (h1). A higher K ratio equals a higher compression over the polymer. Low compression ratios are recommended for amorphous or heat-sensitive materials. Higher compression ratios are recommended for semi-crystalline materials. For example, for POM, K = 2.6:1.

- Plasticizing capacity

  The plasticizing capacity is the machine's capacity if it were acting as an extruder. It can be used to calculate how much time the machine takes to plasticize a kilo of material (unit: kg/h). This capacity is used to compare different machines.

2.4 Screw

The "universal screw" is the most widely used in the injection molding industry. Despite not being a special screw for a specific polymer, it can be used with most polymer materials.

The universal screw has three main zones with different designs and functions. The length of each zone is shown in Figure 2.9 (i. e. in screw diameter terms (Ds), 5D is equal to five times the screw diameter).

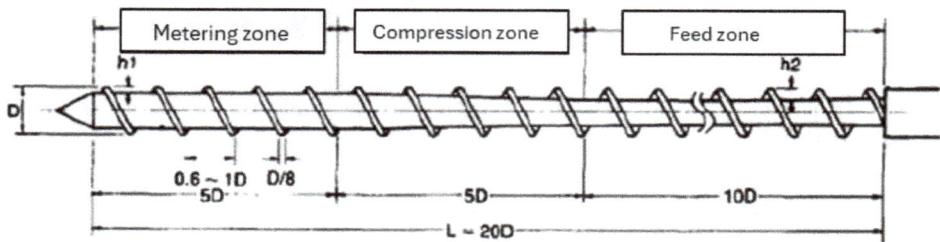

Figure 2.9 Universal screw

Feed zone

This is located at the rear end of the screw, near the hopper. The material, in pellet form, is fed into the injection unit. In this zone, the screw flights are deeper. The solids need good transport properties for transport to the front parts of the screw. The coefficient of friction between the pellets and the screw is important in this zone. Essentially, the injection unit moves solids pellets efficiently forward into the feed zone.

Compression zone

A decrease in the available volume inside the barrel causes progressive compression of the material, heating it up due to friction and shear – here the flight depth is decreasing. This heat contributes to the melting of the material and therefore also causes an increase in the material-specific volume. In this zone, the air between the pellets is displaced toward the feed zone by the compression and does not pass through to the metering zone. It is in this compression zone that we apply compression until 80% of the heat energy needed to melt the polymer is reached (Figure 2.10).

The compression rate is another screw characteristic; this is the ratio between compression ratio (K ratio) and screw compression zone length.

A short screw compression zone promotes faster compression of the material, offering more aggressive compression than a long compression zone.

Metering zone

The melt is mixed and homogenized here.

The ideal size of each of these zones varies with the type of thermoplastic used, especially in the compression zone. For semi-crystalline materials, this zone will be shorter than for amorphous materials.

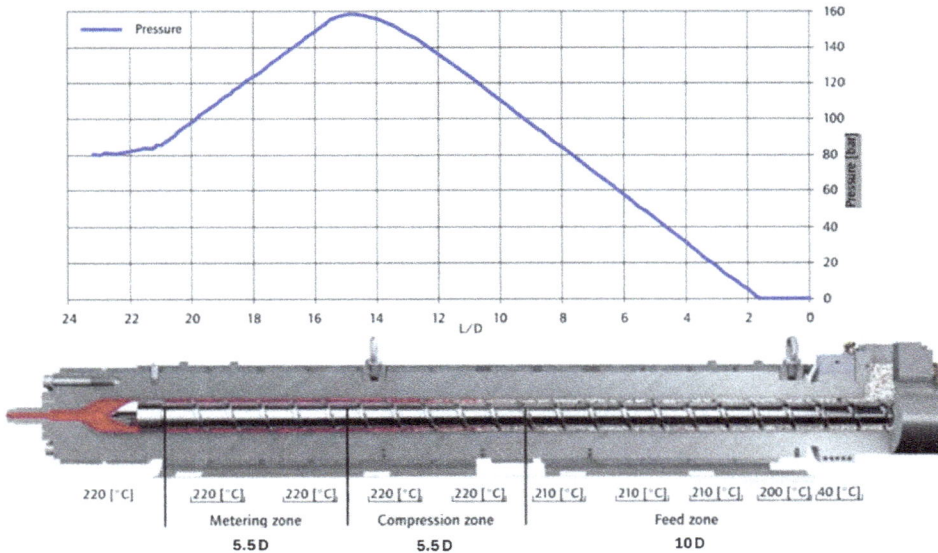

Figure 2.10 Development of internal pressure in the barrel (Source: Wittmann Battenfeld)

The screw is one of the most important elements in an injection molding machine, along with the barrel and the screw valve tip. These elements are essential to the melt quality, process repeatability, and precision of the injection molding process.

At first glance, the screw may seem like a helical coil whose main function is to transport pellets into the barrel to be melted. But this is far from reality. The screw, barrel, and tip valve are designed to feed into the barrel as the pellets travel from the barrel to the injection unit nozzle – the "screw–barrel mode of action".

2.4.1 The Screw–Barrel Mode of Action

Feed zone: This is where filling of the screw's helical channel with solids or pellets takes place. As the machine screws, the barrel pushes and catches the pellets that drop into the slot with each revolution. The performance depends on the screw geometry, the thread angle, the channel width, and the screw speed.

In this zone, the rotating screw exerts a force to push the pellets forward. The friction between the pellets and the screw must be less than the friction against the barrel chamber. Imagine a nut screwed onto the center of a screw. If we turn the screw, it will move, but the nut will not move along the screw. In this case, the nut is the pellets compressed against the barrel and the screw is the machine screw. If we install something to impede the nut's rotation, the friction will be increased and transmitted, and the nut will move along the screw.

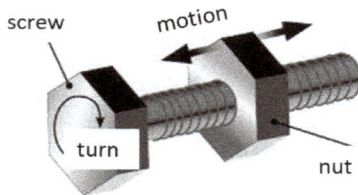

The performance depends on the coefficient of friction between the polymer, the screw, and the barrel. For optimum performance, the pellets must "stick" to the barrel and slide on the screw. We need more friction between the material and the barrel than between the material and the screw.

Compression zone: In this zone, pellets are compressed and fill the channel volume. The pellet-pushing force generated by the screw and the thread angle meets an opposing force which facilitates the back pressure during the metering movement, which also takes place in this zone. Back pressure is essential for the proper functioning of the screw–barrel mode of action because it gives rise to a rotational movement in the screw channel.

2.4.2 The Rotational Movement of the Screw Mode of Action

At 2/3 of the barrel height, the movement and internal speed of the pellets is zero. Above 2/3 of the barrel height, a movement toward the active flank of the screw is generated. Below 2/3 of the barrel height, a movement toward the passive flank is generated.

These movements give rise to a circular mixing movement in the screw channel that is essential for obtaining a homogeneous polymer melt with no unmelted polymer and homogeneous mixing of additives, masterbatches, reinforcements, etc. (Figure 2.11). This is achieved through this circular mixing movement that drives the screw–barrel mode of action (Figure 2.12).

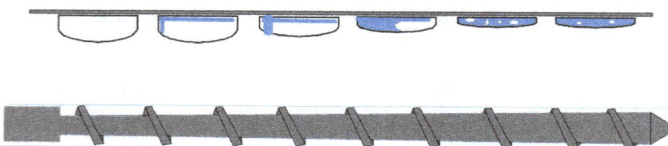

Figure 2.11
Diagram showing the progress of the melt polymer in the screw

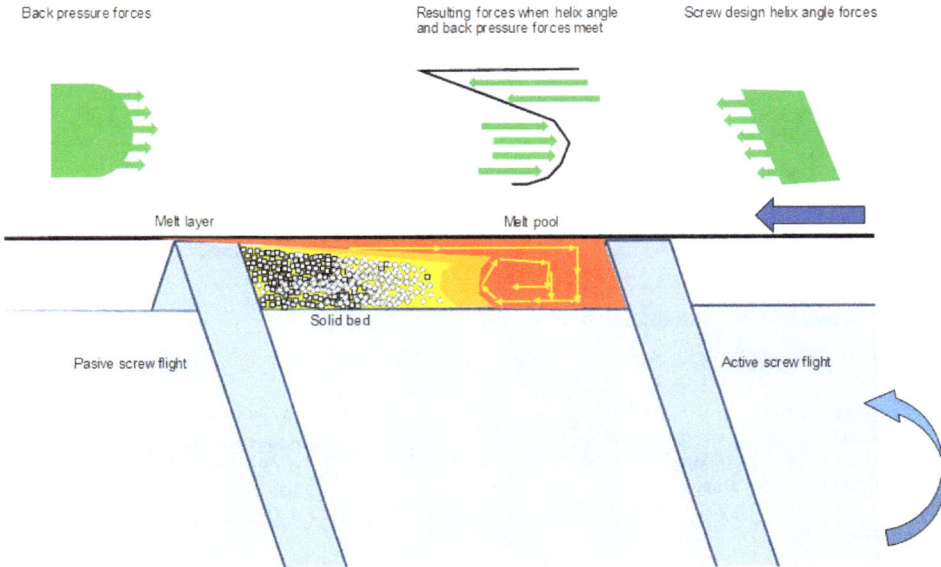

Back pressure forces

Resulting forces when helix angle and back pressure forces meet

Screw design helix angle forces

Melt layer

Melt pool

Solid bed

Pasive screw flight

Active screw flight

Figure 2.12 Screw mode of action

2.5 The Pressure Multiplier

The injection unit acts as a pressure multiplier. The hydraulic pressure applied by the hydraulic pump is multiplied by the intensification ratio (IR) to obtain the specific injection pressure applied to the material.

For example, in Figure 2.13, a hydraulic pressure of 100 bar becomes a specific injection pressure of 1000 bar applied to the material. The intensification ratio in this example is 1:10. This means that a hydraulic pressure of 100 bar applied to a piston area of 100 cm^2 pushes the piston forward with a force of 10,000 kg. If this 10,000 kg force is applied to a screw with an area of 10 cm^2, the pressure applied to the material in front of the screw is 1000 bar. So, a hydraulic pressure of 100 bar becomes a specific injection pressure of 1000 bar when applied to the material.

According to the hydraulic and specific pressure chart (Figure 2.14), if we divide the specific injection pressure by the hydraulic pressure, we will know the ratio of the hydraulic piston area to the screw area. In scientific injection molding, we use the specific injection pressure, not the hydraulic pressure, because the specific injection pressure is the real pressure applied to the material, which can also be essential for transferring the process to or replicating the process in other injection molding machines.

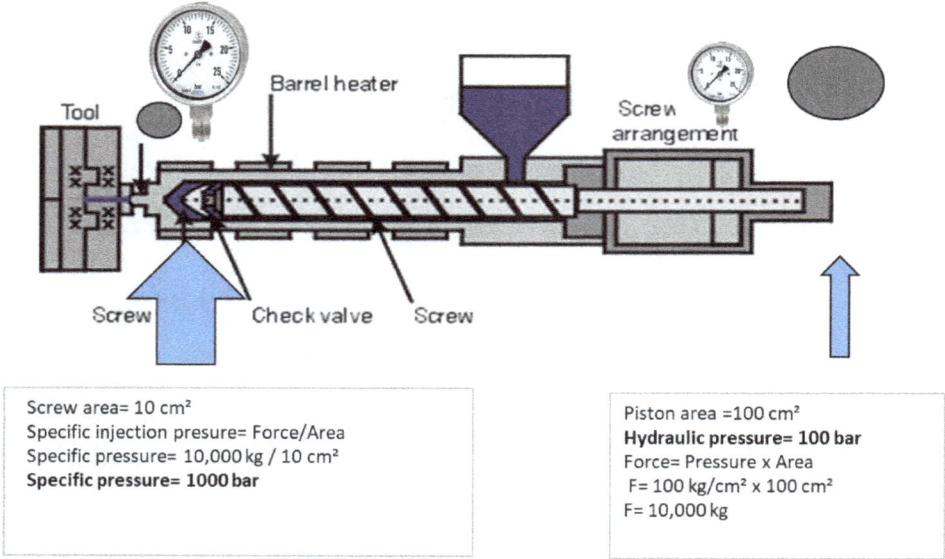

Screw area= 10 cm²
Specific injection presure= Force/Area
Specific pressure= 10,000 kg / 10 cm²
Specific pressure= 1000 bar

Piston area =100 cm²
Hydraulic pressure= 100 bar
Force= Pressure x Area
 F= 100 kg/cm² x 100 cm²
F= 10,000 kg

Figure 2.13 Pressure multiplier diagram

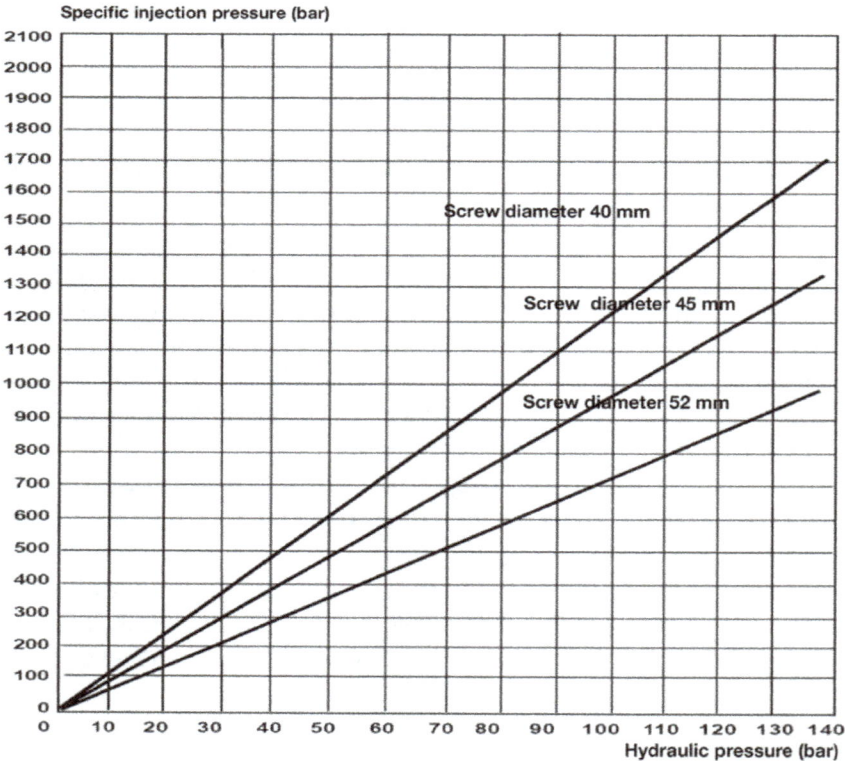

Figure 2.14 Hydraulic pressure versus specific pressure for different screw diameters

Injection Pressure Conversion

We can use the spreadsheet in Figure 2.15 to convert hydraulic pressure to specific pressure and vice versa if we know the machine's intensification ratio. Furthermore, we can use the spreadsheet to calculate the equivalent injection pressure from one injection machine to another with a different intensification ratio in order to replicate the same pressure conditions in both machines.

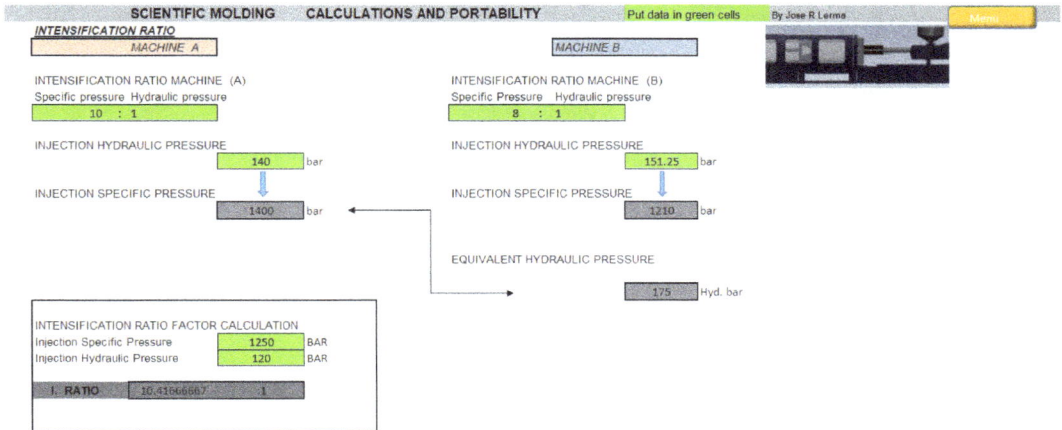

Figure 2.15 Spreadsheet showing hydraulic and specific injection pressure conversions

2.6 The Right Injection Molding Machine

Source:
Coscollola Comercial,
Krauss Maffei

Source:
Wittmann Battenfeld Spain

Source:
Arburg

Figure 2.16 Different injection machine options (Source: Wittmann Battenfeld, Arburg)

When we have a new mold for a new project, or when we need to invest in a new injection molding machine, we must consider some important factors in order to make the best machine selection and choose the right machine characteristics for the mold or project that we are going to develop. We should also follow the raw material manufacturer's recommendations – they understand the material's behavior and performance.

Factors for Proper Machine Selection

- Required melt temperature

  Factors such as recommended melt temperature (beware in the case of high-performance polymers, such as PPSU, PPS, PEEK, etc.)

- L/D ratio

  Some materials are more susceptible to heat and will break down easily and quickly under an inappropriate L/D ratio. Also, a short L/D ratio is a potential root cause for issues with unmelted polymer.

- K ratio (compression ratio)

  Beware, as some materials are so susceptible to compression in the barrel that a maximum K ratio needs to be adhered to. Polymers such as TPU, POM, and PVC can be broken down in the barrel as a result of high compression.

- Recommended clamping tons per unit area

  We can calculate the required clamping force (see Section 2.2.3). With this calculation, we can define the size of the clamping force machine.

- Maximum injection pressure required

This is an important factor in the case of high-viscosity materials or thin parts that require high injection pressure levels. If we do not have enough injection pressure, we will not fill and pack the cavity well and we will probably have a "pressure-limited" process with no injection-speed control and, therefore, an uncontrolled process highly susceptible to material viscosity variability, such as a change in material viscosity.

- Residence time

  All polymers have a maximum residence time at melt temperatures. If the material exceeds this maximum time at these temperatures, the material will break down, lose molecular weight, and the material's properties will drastically decrease. This decomposition depends on both time and temperature. Higher temperatures result in a shorter time before the material breaks down and lower temperatures result in a longer time before the material breaks down.

  Using the spreadsheet in Figure 2.17, we can calculate the material's residence time in the barrel to ensure that the material does not begin to break down. With this spreadsheet, we can calculate the residence time in two different injection machines. For this calculation, we need the following data: screw diameter, maximum metering stroke available, cycle time, and metering stroke.

- Ratio metering stroke- screw diameter

  Furthermore, the residence time should be used to control this ratio to help us choose an appropriately sized injection unit. The recommended ratio is between 1 and 3 (4 at most); see Figure 2.18.

- Optimal ratios: 1D to 3D
- Exceptional ratios: 3D to 4D
- Ratios to avoid: < 1D and > 4D

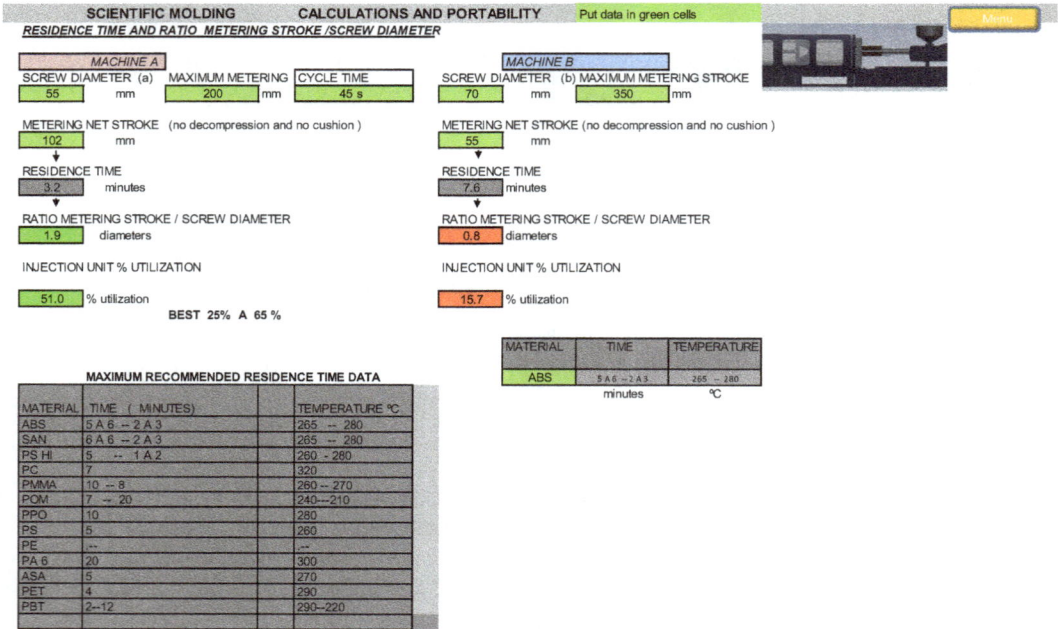

SCIENTIFIC MOLDING	CALCULATIONS AND PORTABILITY	Put data in green cells

RESIDENCE TIME AND RATIO METERING STROKE /SCREW DIAMETER

MACHINE A
SCREW DIAMETER (a) | MAXIMUM METERING | CYCLE TIME
55 mm | 200 mm | 45 s

METERING NET STROKE (no decompression and no cushion)
102 mm

RESIDENCE TIME
3.2 minutes

RATIO METERING STROKE / SCREW DIAMETER
1.9 diameters

INJECTION UNIT % UTILIZATION

51.0 % utilization
BEST 25% A 65 %

MACHINE B
SCREW DIAMETER (b) MAXIMUM METERING STROKE
70 mm | 350 mm

METERING NET STROKE (no decompression and no cushion)
55 mm

RESIDENCE TIME
7.6 minutes

RATIO METERING STROKE / SCREW DIAMETER
0.8 diameters

INJECTION UNIT % UTILIZATION

15.7 % utilization

MATERIAL	TIME	TEMPERATURE
ABS	5 A 6 -- 2 A 3	265 -- 280
	minutes	°C

MAXIMUM RECOMMENDED RESIDENCE TIME DATA

MATERIAL	TIME (MINUTES)	TEMPERATURE °C
ABS	5 A 6 -- 2 A 3	265 -- 280
SAN	6 A 6 -- 2 A 3	265 -- 280
PS HI	5 -- 1 A 2	260 - 280
PC	7	320
PMMA	10 -- 8	260 -- 270
POM	7 -- 20	240--210
PPO	10	280
PS	5	260
PE	--	--
PA 6	20	300
ASA	5	270
PET	4	290
PBT	2--12	290--220

Figure 2.17 Spreadsheet for calculating the residence time

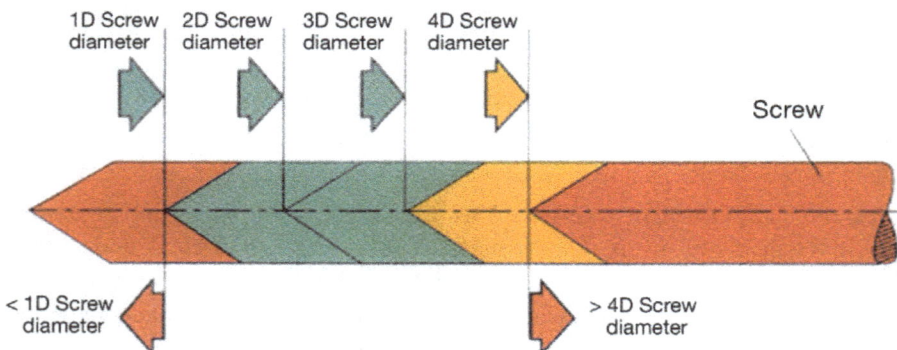

Figure 2.18 Optimal range of injection unit capacity (D = screw diameter). L = Metering stroke. Optimal L/D ratio: 1D–3D; in exceptional cases L/D ratio: 3D–4D; L/D ratios of < 1D or > 4D are not recommended

In order to assign a machine to a specific mold, it is always recommended to do a preliminary estimation of the following:

- Residence time
- L/D ratio. (metering stroke-screw diameter ratio)

> - It is essential to use the most suitable machine for each mold and to do the calculations above.
> - There is no excuse for not doing this!

3 Knowing the Reliability and Performance of Injection Molding Machines

No matter how good the methods, tests, training, and strategies we use, if our machine is not performing well or reacting correctly, we will not have reliability and we will not get better results. That is why it is critical to know the reliability and performance of our machines in order to properly apply scientific injection molding. For example, knowing if our machine is set accurately to apply the predetermined injection speed, and whether or not our machine requires some calibration, will help us make decisions regarding preventive maintenance and recalibration of the injection machine. It is important to be aware of possible wear and tear so that we can keep it under control and decide when it is necessary to intervene.

We will not reproduce a process long after its validation on the same machine if the machine is not in the same condition as when the process was validated. Deviations in injection machine performance and reliability will result in process differences even if the conditions appear to be the same. This is one of the reasons that we often do not understand why a process has deviated from the original one when the machine settings or inputs are exactly the same after having been verified.

Some of the questions we should ask ourselves are: "Do we know if our machine meets the requirements of the predetermined injection speed setting?", "Do we know the Delta P value that we have?", "Is our screw tip accurate and not worn out?", "Is the check ring sealing ok?", "Does the screw accelerate during the injection stroke as usual?"

Some aspects should be analyzed and tested in order that the reliability and performance of all injection machines may be controlled. These include:

- Mold filling time repeatability test
- Load sensitivity test
- Delta P test

- Dynamic and static screw-tip sealing test

- Injection-speed linearity test

- Screw acceleration test

- Pressure response test

After carrying out these tests on the machine, we will be able to determine the performance status and capacity of our machine. Some of them will not tell us if the machine is good enough for accurate production; however, these tests and tools in scientific injection molding will indicate the real situation of our machine to monitor in the event of performance deviations, allowing us to act accordingly. In this chapter, we will describe all of them, how to carry them out and use them, as well as the information they provide us with.

3.1 Mold Filling Time Repeatability Test

This test is used to determine the precision of the injection unit during the mold filling time in each cycle. The mold filling time – the entire injection process's most critical time – is the consequence of the melt flow speed during the filling of the cavities, so its repeatability is very important. Variations in filling time or injection time indicate variations in the molded parts.

Some injection molding authors tell us that the difference in injection time values should not exceed ±0.04 seconds. I prefer to determine these acceptable injection time differences as percentage values, with a maximum correct value for injection time deviations of 0.03 seconds per second of injection time (i.e. a maximum deviation of 3% between shots). For example, the maximum acceptable deviation would be 0.12 seconds for a four-second filling time, and the maximum acceptable deviation would be 0.3 seconds for a ten-second filling time. But for a one-second filling time, the maximum deviation would be 0.03 seconds.

To carry out this test, we run at least eight shots at three different injection speed settings, taking into account the low, medium, and high speeds of the possible injection speed range. We must remember not to limit the injection process by always setting the injection pressure limit above the necessary injection pressure value. From the resulting mold filling time data, we can calculate the range of times, the slowest injection and the fastest injection times as well as the injection time deviation (%). To do this, we can use Figure 3.1 as an example.

SCIENTIFIC MOLDING   CALCULATIONS AND PORTABILITY Put data in green cells   Menu

FILLING TIME REPEATABILITY

METERING STROKE OR VOLUME   50   mm or cm³

	Injection speed fast	Injection speed medium	Injection speed slow
Inj speed (mm/sc ó cm³/sc)	120	50	10
Theoretical injection filling time,sc	0.4	1	5

Filling time , sc from 8 consecitive cycles	Injection speed fast	Injection speed medium	Injection speed slow
1	0.60	2.02	5.00
2	0.50	2.00	5.00
3	0.50	2.00	5.15
4	0.60	2.00	5.00
5	0.50	2.07	5.00
6	0.40	2.00	5.00
7	0.50	2.00	5.00
8	0.50	2.00	5.00
Minimum injection speed filling time	0.60	2.07	5.15
Maximum injection speed filling time	0.40	2.00	5.00
Filling time range	0.20	0.07	0.15
Filling time deviation %	33.3	3.3	2.9

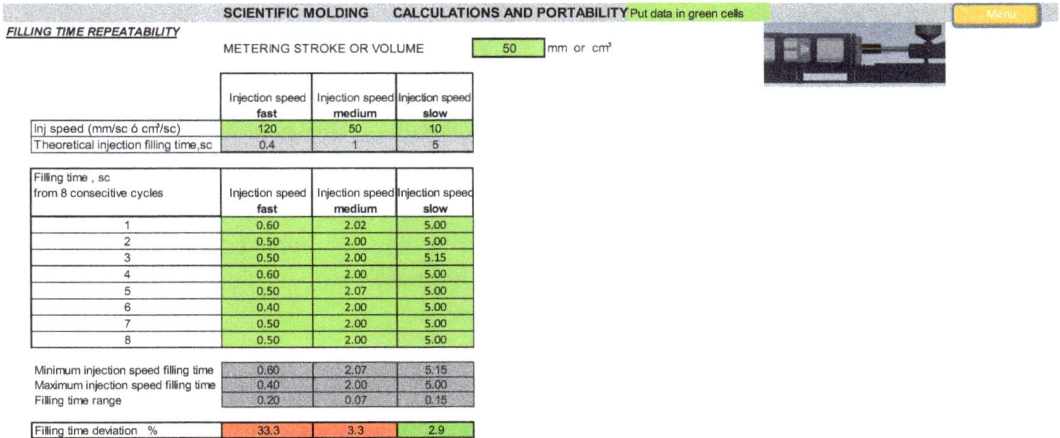

Figure 3.1 Spreadsheet for determining the precision of mold-filling times

This test will indicate the precision with which the machine repeats the injection time and deviation data at different possible injection speed ranges. The smaller the deviation in times, the more precise our injection machines will be. Furthermore, the data obtained will serve as a reference for subsequent tests to be able to assess whether there are increases in the injection time deviation (%) so that, if necessary, we can calibrate the injection machine speed to return to the machine's usual values.

3.2 Load Sensitivity Test

This test reports the injection machine's ability to maintain the filling time under different loads; for example, differences in material viscosity due to different batches of raw material, a different percentage of regrind, changes in material color, changes in melting temperature, different mold temperatures, and different additives. This test helps us to assess the injection machine's performance in relation to how the machine is able to adapt or compensate for the injection pressure according to the load or resistance offered by the material to flow (i. e. the real viscosity of the melt material).

In this test, we compare the injection pressure values reached for the injection unit to fill up the mold from the metering position to the V/P switchover point and the mold filling time to reach the V/P switchover point, i. e., the dynamic filling stage, under two different conditions:

- Mold filling under usual process conditions
- No-melt material filling condition or no-load filling condition

With these values, we calculate the percentage of load sensitivity compensation made by the injection machine for every 100 bar of injection pressure required.

3.2.1 Machine Configuration for Carrying Out the Pressure Compensation Test

- Set the injection machine to inject by switching to the holding pressure stage on the basis of stroke position or volume, in mm or cm^3.

- Set data for the mold filling speed, obtained from the relative viscosity test or the in-mold rheology test. Here, it is possible to set the standard injection speed used during the production of series parts. Alternatively, the test could be also carried out at two or three different injection speeds.

- Usual V/P switchover point during the production of normal series parts.

- Ensure enough available injection pressure and Delta P, making sure not to limit the injection pressure process.

- Lower the holding pressure to ensure we avoid sink marks or even slightly empty parts.

- Set the minimum holding pressure time or leave it at zero. In this test, we are not interested in the packing and holding phase – only the filling phase will be analyzed.

- Start the injection shot and refine the V/P switchover point to get parts at 90–98% of the total filling.

Record two outputs – the filling time and filling pressure at the switchover point. Here, we will have the values for:

> Injection pressure (PK1)
> Injection time (FT1)

In order to adhere to the predetermined injection speed setting and repeat the mold filling time (the process's most critical time), the injection machine must apply the necessary injection pressure in each injection cycle. Depending on certain factors (e. g. the viscosity of the material, mold and melt temperatures), this pressure may vary because the injection machine automatically carries out an injection pressure "compensation" according to the load. The ability to self-compensate for these pressure differences is an important injection machine feature, which must be controlled as it will differentiate accurate injectors from those that are not.

In the test, two opposing conditions are compared: one with the molten polymer under the aforementioned usual injection molding conditions and, the other, injecting without melted material in front of the screw – this is called the "no-load condition".

This "no-load condition" injection test with no-melt material is often complicated to carry out by purging in a manual machine mode, as modern injection machines currently work in safe mode with low injection pressures and low speeds when they are in manual mode and it is not possible to automatically replicate the injection conditions. Therefore, the decompression movement is used for these low-hold injection shots or under a no-melt material condition. With this linear movement, the screw recovery position will be the same as during the normal metering stage, with a stroke equivalent to the metered amount required, so we will get the screw to the position of "metered amount reached" and be able to execute an injection shot cycle without material in the front area of the screw (i. e. while it is "empty", or unloaded). We will carry out an automatic or semi-automatic shot, recording the injection pressure and mold filling time data for this no-melt material or "no-load condition".

Record two outputs – the filling time and filling pressure at this point. Here, we will have the values for:

> Injection pressure (PK2)
> Injection time (FT2)

The mold filling time should be the same in the normal injection melt material condition as in the no-melt material condition (i. e. FT2 = FT1) – what changes is the injection pressure needed to execute the injection stroke, which should be much lower in the case of PK2.

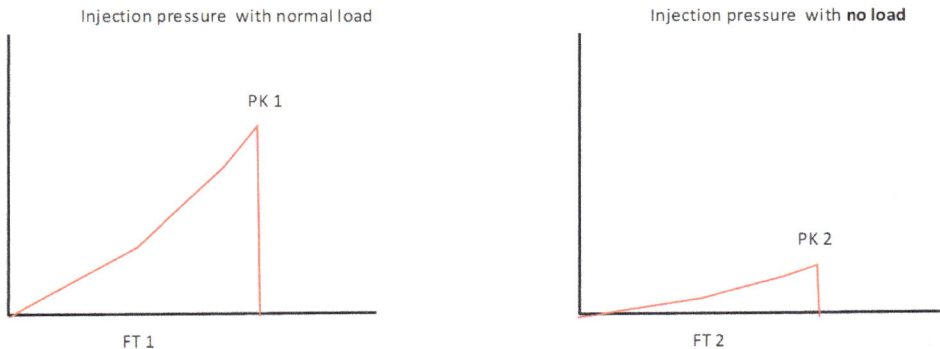

To calculate this load sensitivity and pressure compensation function in the injection machine, we should use the following formula in the load sensitivity test:

$$((FT1 - FT2) \div FT1) \div ((PK1 - PK2) \div 1000) \times 100$$

Results greater than 10% are not acceptable. In that event, the injection machine should be calibrated. Values close to zero or slightly negative are acceptable.

Carrying out this test at three different injection speeds (slow, medium, and fast) is recommended. If your machine passes the test at all three speeds, it means that it is reliable.

3.2.2 Spreadsheet for the Load Sensitivity Test

In this spreadsheet (Figure 3.2), by filling the green cells with the data obtained from the test, we can calculate the machine's load sensitivity or load compensation (%) for the injection pressure per 100 bar. The injection pressure and mold filling time data in both normal and no-load or no-melt material conditions should be recorded in the spreadsheet, and pressure and time data could be the average data obtained from at least ten injection shots.

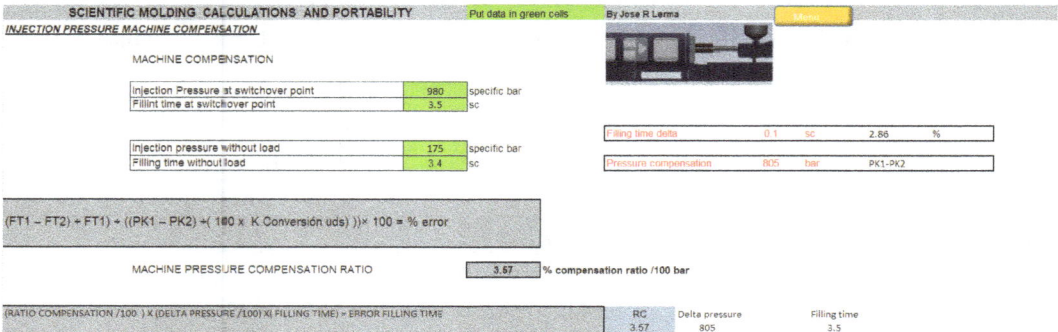

Figure 3.2 Spreadsheet for determining machine compensation of injection pressure

The values resulting from the load sensitivity or load compensation ratio obtained should be understood as follows:

- Up to 5%: correct (machine is precise)

- Between 5% and 10%: acceptable for injection molding with low tolerance requirements – it is necessary to review for molding parts with tight tolerances

- More than 10%: machine is not precise or consistent – check the injection molding machine

Other Tests for Injection Pressure Compensation Analysis

Another way to change the load during the dynamic injection stroke is to promote a change in melt material viscosity and therefore intentionally change the load on the injection machine to test the injection machine's reaction, behavior, and reliability. For example, by modifying the viscosity of the melt material, we can interrupt a machine cycle when a certain level of material degradation has been reached within the injection barrel, after we automatically inject some shots. We can determine how many injections are needed for the injection machine to return to normal pressure and filling time values and we will be able to check how the injection machine self-regulates and compensates by decreasing the viscosity of the material due to the effect of the higher residence time.

In the same way, we can test the load sensitivity by changing important melt temperature settings (e. g. by reducing the melt temperature to increase the viscosity of the melt material by checking the injection machine reaction and output changes, and by checking how the pressure is self-compensated to maintain the injection time). This will test the machine's reaction to batch changes, regrind material (%), different-colored masterbatches, etc.

3.3 Delta P Test

This test is used to obtain the correct Delta P value in our injection machines and for setting the right Delta P data. This well-applied Delta P data will provide us with a more robust and consistent process with less variability. The definition of Delta P is the difference between the real injection pressure needed for filling the mold and the injection pressure limit setting on the machine.

Allegory of a car's cruise speed control:

To engage cruise control in a car, you need to have more horsepower (CV or HP) available than you are currently using. If you are driving up a steep hill with the car in cruise control, the car will need enough horsepower to maintain the desired speed. On the other hand, if you drive down a slope in cruise control, the car will need less horsepower to maintain the same speed. It will be the cruise control system that regulates the horsepower needed to meet the speed set in cruise control.

Figure 3.3 The cruise control in a car

The same concept applies to an injection machine during filling control in the dynamic cavity-filling phase. The car's horsepower would be the equivalent in the injection machine to the available injection pressure. During the filling phase, we must set – as a control measure at this stage – the injection speed (just as we control the speed and not the horsepower of a car) and not the injection pressure (that will be an output, a consequence of the injection pressure needed to meet the mold-filling speed set on the machine control).

If the injection speed and therefore the filling or injection time is the most critical time of the process, meaning the injection machine must reproduce it precisely, we are talking about a maximum acceptable deviation in injection times between cycles of ±0.04 seconds, or 3%. How can the injection machine meet these time-repeatability requirements if we are going to process material batches with different viscosities, different regrind, and so on? The machine has only one resource to do this, which is to have enough pressure in reserve to self-regulate (with the available pressure) the different conditions found during the different injection shots with the mentioned variabilities: material viscosity, regrind, additives, etc.

The predetermined injection pressure limit should be high enough to allow the injection machine to be able to deliver the established speed. Remember that delivering the established injection speed is one of the injection machine's most important goals. That is why we need to set the injection pressure higher than the pressure needed to fill the cavities. This way, we will have injection pressure available on reserve. Otherwise, depending on which values we have established for this injection pressure limit, two undesirable situations may occur:

- The injection pressure limit is set lower than the injection pressure needed to reach the predetermined injection speed.

  This situation is referred to as a **"limited injection pressure process"**. This is a process to avoid, as we lose control of the filling speed when we operate with limited pressure. Unfortunately, this is a method that some (even large) companies use as normal injection pressure setting conditions. It will be also an injection molding process which is not self-regulated at all, based on variable batch viscosities, using regrind, masterbatches, and other aspects of the process that affect the filling of the cavity and, consequently, the injection pressure. A "limited injection pressure process" will definitely reduce the injection speed when the injection pressure needed is close to the injection pressure limit. This process will lack robustness, consistency, and repeatability.

- The injection pressure limit is set at an excessively high level.

  In this situation, we will not have a limited injection pressure process as was the case in the previous situation mentioned, but we will have a process that can cause damage to the mold, cavities, machine, etc.

If we look at the case of a multi-cavity mold, there is a possibility that a cavity may have accidentally caused problems during the filling phase and the machine could try to fill the entire volume of the mold. In this case, with such a high level of injection pressure available, the highest levels of injection pressure would be reached, which can cause damage to the mold, machine, etc.

Therefore, we need to set the injection pressure limit parameter above the pressure level needed to fill the mold, so we have extra reserve pressure that the machine will use in case of viscosity changes, etc. This extra pressure is what we refer to as Delta P.

Whether your machine is electric or hydraulic, closed-loop or open-loop, your goal must be to maintain a constant filling time, or you must have a certain amount of additional pressure available. This is the Delta P – the necessary pressure difference between the pressure available in the machine and the pressure needed to fill the mold.

A suitable industrial value for Delta P is a hydraulic injection pressure of 15–30 bar, or a specific injection pressure of 150–300 bar. However, you can reduce this Delta P value for sensitive molds or if your molded part precision requirements are that high.

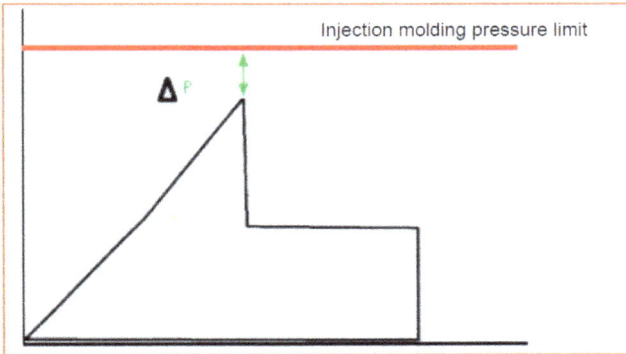

Figure 3.4
Graph showing Delta P

Delta P Determination Procedure

To determine a particular machine's Delta P value, it is necessary to inject a mold that requires an injection pressure higher than 60% of the maximum pressure available in the machine.

Previously:

- We should be able to control the maximum pressure peak (bar) of each shot whilst filling the mold during the first injection phase, just in the V/P switchover point.

- Ensure that we can also monitor the mold filling time, from the beginning of the injection stroke until the V/P switchover point is reached.

Machine configuration:

1. Switching the system to the holding pressure phase must proceed via the screw position or volume.

2. The machine's injection speed could be selected from a previous in-mold rheology test or the injection speed set during the molding process on a part production series.

3. Ensure that the holding pressure phase is not activated (i. e. set the holding pressure and holding pressure to zero).

4. Check that there are no problems with short shots or short parts (e. g. with ejectors, slides, etc).

5. Adjust the injected volume until the desired V/P switchover point is obtained.

6. At this point, compare the peak injection pressure with the machine's predetermined injection pressure limit.

When comparing the injection pressure needed to reach the V/P switchover point against the injection pressure limit set on the machine, we can check these two possible situations regarding the injection pressure:

- Equal or very similar to the injection pressure limit set on the machine

- Lower than the injection pressure limit set on the machine

Situation I: The peak injection pressure at V/P switchover point is equal or similar to the injection pressure limit. We have no Delta P, and we are probably molding with limited pressure (limited injection pressure process).

To carry out this Delta P test, we should enter the injection pressure reached, pressure limit (required pressure and set pressure limit) and filling time data directly into Table 3.1. After increasing the injection pressure limit in 100-bar steps (specific pressure) or 10-bar steps (hydraulic pressure), the pressure and filling time data should be entered into the table.

During the test, once an adequate injection pressure limit is set, the machine will repeat the filling time over and over. At this time, the process is not limited.

If you use the Delta P test spreadsheet (Figure 3.5), you can see that the Delta P graph takes on a "flat" shape in the filling time data (red line).

Table 3.1 Delta P Test

Shot	Pressure Limit (bar)	Filling Time (s)	Injection Pressure Reached (bar)	Delta P (bar)
1	1100	2.4	1050	Unacceptable
2	1200	2.2	1130	Unacceptable
3	1300	1.9	1270	Unacceptable
4	1400	1.85	1380	Unacceptable
5	1500	1.7	1410	Unacceptable
6	1600	1.65	1550	Unacceptable
7	1700	1.6	1575	Unacceptable
8	1800	1.6	1570	Acceptable repeat filling time
9	1900	1.6	1580	Acceptable
10	2000	1.6	1570	Acceptable

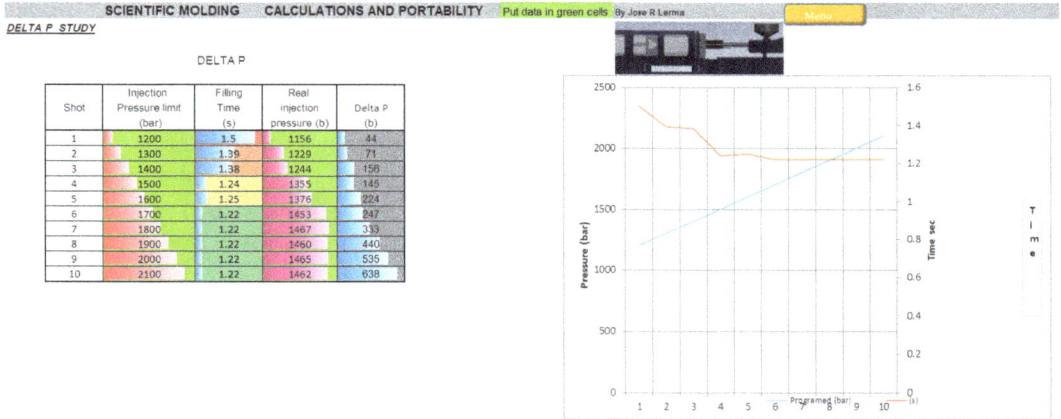

Figure 3.5 Delta P test spreadsheet

Situation II: The injection pressure at V/P switchover point is lower than the injection pressure limit. In this case, we are not in a limited injection pressure process, so we must reduce the injection pressure until we notice that the filling time increases. At this precise moment, we have limited the injection process.

Just above this injection pressure value, we should increase the injection pressure limit set to 15–30 bar (hydraulic pressure) or 150–300 bar (specific pressure). In this case, we enter the data for the injection pressure required to fill the mold, the injection limit setting, and the filling time into Table 3.1.

To reduce the injection pressure limit, set it in 100-bar (specific pressure) or 10-bar (hydraulic pressure) steps and enter the data in each step. When the mold filling time increases slightly, this is the injection level at which we have limited the injection pressure.

Above this point, we should set an injection pressure limit of 15–30 bar (hydraulic pressure) or 150–300 bar (specific pressure). This is the correct Delta P value.

There are some injection machines that, before the injection pressure needed for filling the mold reaches the injection pressure limit, already cut the injection pressure slightly (via software and hydraulic system design). It is therefore necessary to do this Delta P test to check the point of pressure at which the machine cuts the injection pressure and reaches a limited process.

3.4 Screw Tip/Check Ring Valve Sealing Test

The mechanical element that, during the injection phase, prevents the molten plastic from returning to the rear of the injection unit and allows the material to advance toward the front area of the barrel during the metering stage is the screw tip (also known as the check ring valve; see Figure 3.6). It is also an anti-return valve that must contain the molten plastic for a good, repeatable injection process. This element must be airtight during the application of injection pressure when the screw moves forward – otherwise material leakage will occur, which will result in significant differences between shots, no repeatability, no robust process, and increased scrap.

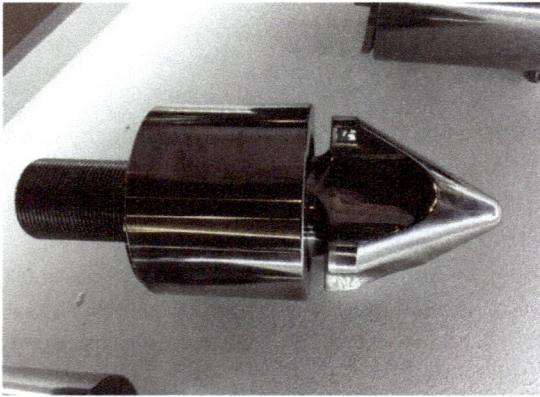

Figure 3.6
Screw tip/check ring valve
example

Problems that may arise with a screw tip/check ring which is worn out, unsealed, or in poor condition:

- Irregular cushion between shots (no stability)
- Irregular metering times
- Excessive shrinkage of parts
- Black spots burned into the material
- Streaks on the part's surface
- Instability of the part's dimensions
- Irregular part weight

This injection unit element should be carefully monitored within the scheduled maintenance to detect wear and tear or lack of sealing. Two types of tests that can be carried out to check the sealing of this element are:

- Dynamic

- Static

The weight of the molded part is an indicator of the sealing of the anti-return valve or check ring valve. Depending on the sector or application of the molded part and the tolerances required in the parts, it is necessary to determine the molded parts' weight dispersion limits (%).

3.4.1 Dynamic Test

This test is carried out at 90–95% of total cavity mold filling.

Injection Machine Configuration for Carrying Out the Screw Tip/Check Ring Valve Sealing Test

Decouple or disconnect the holding pressure phase. Set the holding pressure time and holding pressure to zero, because the behavior of the screw tip seal that we are going to test is mainly exhibited during the volumetric filling phase or dynamic phase, not during the holding pressure phase.

This test will involve the execution of ten injection shots, whose weight will be controlled. 90–95% of the cavities will be filled with the holding pressure phase set to zero (pressure and time) to exclusively analyze the filling phase.

Start the process and once stabilized, record ten injection shots and check the weight obtained for each shot. If the screw tip is airtight and there is no wear and tear on the injection barrel, the weights will be consistent.

We can use the spreadsheet in Figure 3.7 to calculate data such as the maximum and minimum weight, variance, or deviation between shots to check and maintain the precision and repeatability of the injection process. In the case of the suggested spreadsheet, condition values such as average, median, range, and variability are calculated. As a reference, a variance greater than 3% indicates that the check ring valve should be replaced or disassembled and inspected.

SCIENTIFIC MOLDING   CALCULATIONS AND PORTABILITY   Put data in green cells   By Jose R Lerma   Menu

SCREW TIP HERMETICITY STUDY

Shot	Shot weight (g)	
1	119.4	g
2	119.6	g
3	119	g
4	119.4	g
5	110	g
6	119	g
7	117	g
8	118	g
9	116.9	g
10	114	g

Total	1172.30	g
Maximum weight	119.60	g
Minimum weight	110.00	g
Average	117.23	g
Median	118.50	g
Range	9.60	g

Variance	8.03	%

Maxim acceptable	< 3%	%

Figure 3.7 Spreadsheet for the screw tip/check ring valve sealing test

3.4.2 Static Test

This test is carried out with short screw movements during the holding pressure stage.

3.4.2.1 Injection Machine Configuration for Carrying Out the Screw Tip/Check Ring Valve Sealing Test

Do not carry out this test on hot runner molds at high temperatures.

The purpose of this test is to analyze and verify the screw tip sealing or barrel wear.

Machine Configuration for Static Screw Tip/Check Ring Valve Sealing Tests and Barrel Wear Tests

- Allow the injection shot and its corresponding distribution channel or runner to cool down inside the mold so that there is no possibility of material entering the mold

- Set the holding pressure to a typical setting for the process, but at least 40–50% of the maximum holding pressure available on the machine settings

- Set the holding pressure time to approximately 10–20 seconds

- Set the metering stroke to approximately 80% of the maximum screw volume

- Set the V/P switchover point by stroke or volume

- Set a V/P switchover point to change to the holding pressure phase **very close to the metering stroke position** reached, with the aim of immediately switching to the holding pressure phase

- Inject a shot and check whether the screw advances or remains static after the V/P switchover point and in the holding pressure phase during the holding pressure time

- Bear in mind that the plastic will be molten and you should only see the screw advance slightly at the beginning of the injection movement due to the compressibility of the molten polymer – the screw should then stop and remain static

- Carry out the same test with different metered quantities, strokes, or barrel capacities (i. e. with 10%, 25%, 50%, and 80% of the total available capacity) – so, check at least three or four different metering strokes from the total metering stroke available on the machine setting

If the test is acceptable at 80% of the total capacity, but the tests at 10% or 50% of the total capacity are not acceptable, you may conclude that the screw tip and the seal ring wear are correct and that the barrel may be worn out.

If the screw advances slowly during testing in different metering positions, the screw tip should be immediately checked or replaced.

3.4.2.2 What about Hot Runners?

This static test cannot be directly carried out with hot runners. Please bear in mind that one condition for this test is not to inject material into the mold.

The test can be carried out by lowering the hot runner setting temperatures until the material cannot enter the mold – alternatively, there are companies that use injection nozzles with closed outlets to carry out the static test. These kinds of nozzles allow us to activate the screw injection movement and execute a pressure injection without the risk of forcing material into the hot runner since this material cannot flow through the nozzle because it does not have an open outlet for the material.

3.5 Injection-Speed Linearity Test

The injection speed is the most critical speed of the entire injection molding process. Consequently, the mold filling time is also the process's most critical time.

Two aspects of the injection machine are very important: on the one hand, the execution of a repeatable and consistent mold filling time (this is, to repeat each cycle at the same injection speed); on the other hand, delivering the machine's predetermined injection speed. Regarding the repeatable injection time, there are authors who propose a maximum dispersion of injection times in the order of 0.04 seconds between cycles or a maximum of 3%. This dispersion is not very wide and so our injection machine must be properly maintained and calibrated in order to meet this requirement. As for the injection molding machine's compliance with the predetermined speed, it is essential that the machine be accurate enough to be able to reach the injection speed setting.

Molded plastic parts and plastic injection molding have increasingly demanding requirements, both in terms of their aesthetics and dimensions. The percentage of scrap that can be produced as a result of the process is reduced to improve competitiveness and productivity. Consequently, our machines and our processes must be more repeatable, robust, consistent, and productive.

If we use an injection speed of 100 mm/s and our injection stroke, from the metering position to the V/P switchover point, is 100 mm, we should expect an injection time of one second. Likewise, if we use an injection speed of 50 mm/s with the same metering stroke, we should expect a filling time of two seconds. But when you carry out these tests on your injection molding machine, this does not always turn out to be the case. If we carry out these simple tests on our machine, we should not be surprised that the actual times are not what we expected.

We can carry out an injection-speed linearity test on the injection machine. This test should be carried out regularly to determine the machine's precision level.

We need to know if our machine – when we set a certain speed – complies with that speed, and, most importantly, we must know what percentage of deviation the machine has regarding the predetermined injection speed. By monitoring the percentage of compliance with the predetermined speed, comparing it with the actual speed, we can monitor the progress of the injection molding machine and be able to decide when the injection molding machine's injection speed should be calibrated, based on scientific evidence and not arbitrarily or empirically.

Injection-Speed Linearity Test Procedure

To carry out the test, we must know the dynamic stroke data – this is the stroke where the screw will travel from the metering position to the V/P switchover point. By subtracting the V/P switchover point position from the metering stroke position, we will obtain the "net dynamic injection stroke".

This test should be carried out with a mold that requires at least 60–70% of the injection molding machine's total metering capacity in order to obtain more real data regarding the machine's speed linearity.

Injection Machine Configuration for the Injection-Speed Linearity Test

- Select the holding pressure switchover system by stroke or screw position by volume
- Set the holding pressure to zero
- Set the holding pressure time to zero (ensure that parts can be removed, even if they come out short)

- Set the injection pressure limit to the maximum pressure available on the machine, to ensure that the process will not be a "limited injection pressure process"

- Adjust the V/P switchover point setting to fill approximately 80–85% of the total part volume (no more than that, because, if we adjust the V/P switchover point setting to fill close to 100% of the volume, when we execute injection shots at high speeds, flash may occur – a situation which should be avoided)

- Inject ten different shots, setting injection speeds from slow to fast, depending on the machine's available capacity – the increments in each shot's injection speed should be kept equal

For each of those ten shots, the following data must be recorded (in addition to the set speed and the shot number):

- Filling time (seconds)

- Injection pressure (bar) – maximum injection pressure reached in the V/P switchover point to holding pressure

By entering this data into the spreadsheet shown in Figure 3.8, we can obtain the actual injection speed reached during filling of the mold, and the percentage compliance in relation to the set speed. We can also see in Figure 3.8 a comparative graph showing the progress of the set injection speed and the real injection speed results taken from the different shots.

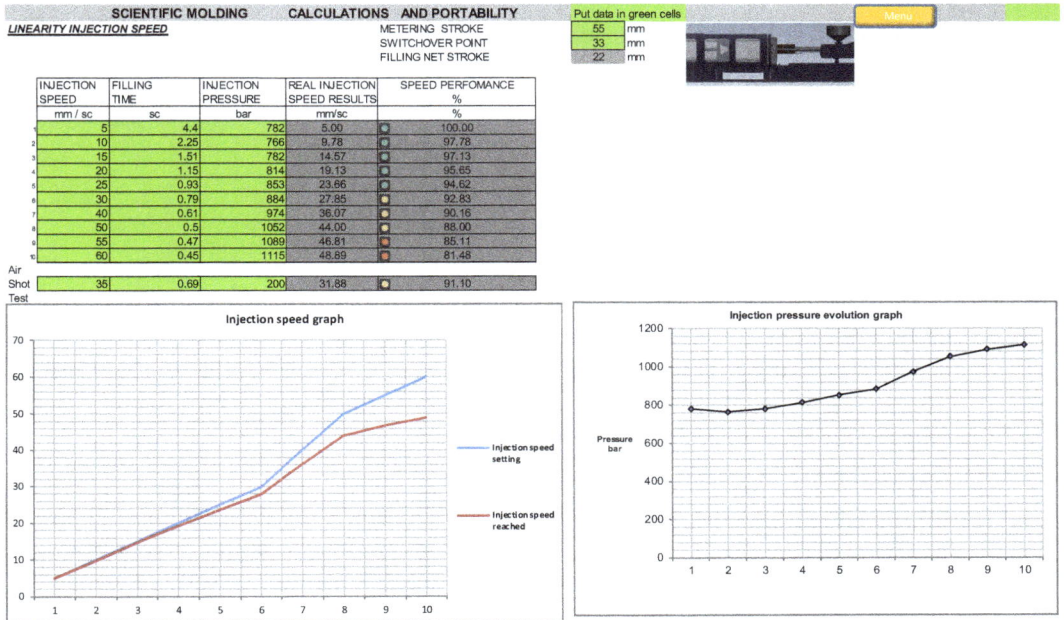

SCIENTIFIC MOLDING   CALCULATIONS AND PORTABILITY

LINEARITY INJECTION SPEED

METERING STROKE	55	mm
SWITCHOVER POINT	33	mm
FILLING NET STROKE	22	mm

INJECTION SPEED	FILLING TIME	INJECTION PRESSURE	REAL INJECTION SPEED RESULTS	SPEED PERFOMANCE
mm / sc	sc	bar	mm/sc	%
5	4.4	782	5.00	100.00
10	2.25	766	9.78	97.78
15	1.51	782	14.57	97.13
20	1.15	814	19.13	95.65
25	0.93	853	23.66	94.62
30	0.79	884	27.85	92.83
40	0.61	974	36.07	90.16
50	0.5	1052	44.00	88.00
55	0.47	1089	46.81	85.11
60	0.45	1115	48.89	81.48

Air Shot Test

| 35 | 0.69 | 200 | 31.88 | 91.10 |

Figure 3.8 Spreadsheet for the injection-speed linearity test

There is clearly always a "gap" between the set injection speed and the real injection speed reached, especially at high injection speeds. This is mainly because the machine needs some time to accelerate the screw from the metering position, where the screw is stopped, until the predetermined speed is reached. The machine needs some time to accelerate the screw in the metering stationary position from an injection speed of 0 mm/s to the injection speed reached. This "gap" will therefore depend entirely on the machine acceleration capacity and its fine control and calibration.

The acceptable speed compliance percentage at high injection rates are those values in the range of 85–90% of the predetermined speed. In other words, the machine has a maximum of 10–15% of the dynamic injection stroke to reach the requested injection speed.

The data obtained, with a larger or smaller gap between the speed reached and speed set, helps us to determine the percentage of machine injection speed compliance compared with the injection speed setting. It also serves as a way to monitor and follow up on the progress of the machine's behavior (regarding the critical parameters of injection speed and mold filling time) over time.

Different speed linearity tests from the same machine can be compared, to enable the monitoring of the injection molding machine's progress and detect deviations in advance. When different injection-speed linearity tests from the same machine are compared and the "gap" between the injection speed reached and the injection speed set increases, the machine should be checked urgently and the injection speed should be calibrated to return to the usual deviations for that particular machine.

If the machine behavior changes with respect to injection speed, key parameters will change and the process will deviate from the standard.

3.6 Screw Acceleration Test

Due to the great importance of the injection speed in the injection process and the relation between shear rate and polymer viscosity (see Section 4.3.2), when we set a certain injection speed, we expect our injection machine to reach the injection speed exactly as per the data to ensure the robustness of the process and the properties of the parts. Additionally, we expect that this injection speed will be repeatable and consistent from cycle to cycle and batch to batch.

As you can see in Section 3.5, the injection-speed linearity test, in which we compare the level of compliance between the set injection speed and the real injection speed reached, there is always a "gap" between the speed requested and the real speed at high injection speeds. As explained in Section 3.5, this effect is directly related to the machine screw acceleration capacity and must be taken into account.

The screw will move at a high injection speed starting from the metering position. To reach the set injection speed, the machine needs some time to accelerate. The screw will finally reach the set speed, but it is logical that this will take a while depending on each injection machine and its condition.

This time spent accelerating the screw means a delay and, therefore, a speed deviation. If we only consider the total injection stroke during the filling movement and the total filling time spent, we can calculate the real speed reached, which will never be equal to the set speed. It will always be less due to unavoidable time required to accelerate the screw movement.

That is why it is important to analyze, check, and monitor how our injection machine accelerates to the predetermined injection speed and if the machine can fully reach it once the acceleration has begun. How long it has taken to reach the speed requested – and the distance traveled to reach it – are also important factors.

To become familiar with the injection machine's behavior, we must know its screw acceleration curve during injection (Figure 3.9). This will give us important information about how our machine behaves and when this test is performed regularly it will enable us to detect deviations in the screw acceleration. Additionally, it will help us understand why some machines, even from the same machine manufacturer, do not behave exactly the same when we set the same injection parameters.

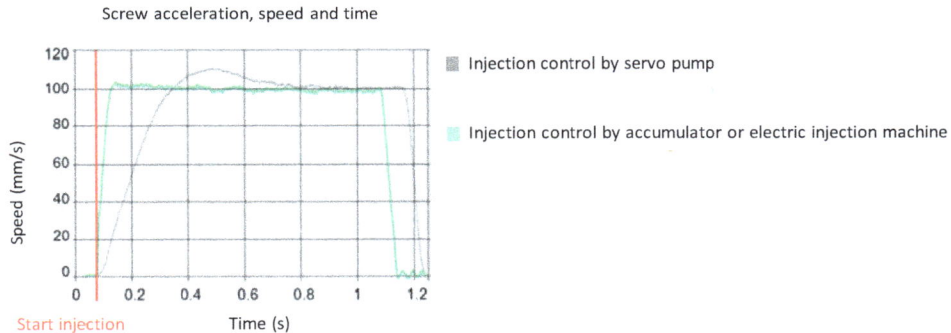

Figure 3.9 Screw acceleration. Source: Arburg

Screw Acceleration Test Procedure

To carry out this test, the metering stroke reached by the screw must be known. We will then sequentially modify the V/P switchover point to the holding pressure point to execute different injection strokes by controlling the mold filling time or the time that the injection machine takes to reach the V/P switchover point from the metering position.

Injection Machine Configuration for the Screw Acceleration Test

- Select the holding pressure switchover system by stroke or screw position by volume

- Set the holding pressure to zero

- Set the holding pressure time to zero (ensure that parts can be removed easily, even if they come out short)

- Set the injection pressure limit to the maximum that the machine allows to ensure that the process will not be a "limited injection pressure process"

- Set the injection speed requested for the test (medium-high speed to see acceleration at certain speeds)

- Set the switchover point to ten different positions, so that more material fills the mold with each shot

- Inject shots at these different V/P switchover points

- We should have at least ten shots recorded with their filling time values (s)

In the spreadsheet (Figure 3.10), record the metering stroke and injection speed at which we will carry out the test. The spreadsheet will propose ten equidistant switchover positions to us. The switchover proposed by the spreadsheet is calculated by dividing the total metering stroke position by ten. By executing the proposed injection strokes from the metering position to the different switchover points and recording the injection time spent for each one, we can calculate the real speed reached at each injection shot and the real speed obtained cumulatively and over the total injection stroke.

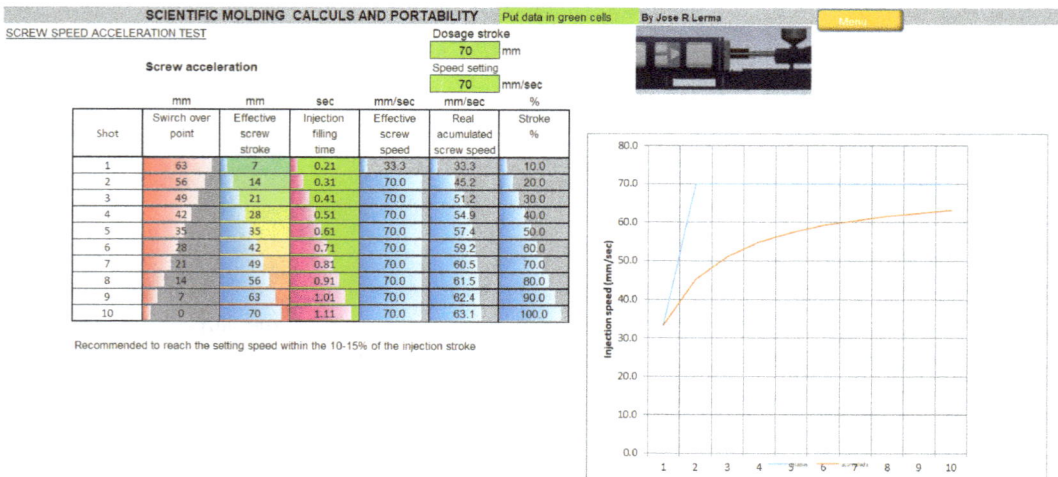

SCIENTIFIC MOLDING CALCULS AND PORTABILITY   Put data in green cells   By Jose R Lerma

SCREW SPEED ACCELERATION TEST

Screw acceleration

Dosage stroke: 70 mm
Speed setting: 70 mm/sec

Shot	Swirch over point (mm)	Effective screw stroke (mm)	Injection filling time (sec)	Effective screw speed (mm/sec)	Real acumulated screw speed (mm/sec)	Stroke %
1	63	7	0.21	33.3	33.3	10.0
2	56	14	0.31	70.0	45.2	20.0
3	49	21	0.41	70.0	51.2	30.0
4	42	28	0.51	70.0	54.9	40.0
5	35	35	0.61	70.0	57.4	50.0
6	28	42	0.71	70.0	59.2	60.0
7	21	49	0.81	70.0	60.5	70.0
8	14	56	0.91	70.0	61.5	80.0
9	7	63	1.01	70.0	62.4	90.0
10	0	70	1.11	70.0	63.1	100.0

Recommended to reach the setting speed within the 10-15% of the injection stroke

Figure 3.10 Spreadsheet for the screw acceleration test

With this test, we will be able to evaluate the injection machine acceleration ramp – how long it takes the injection machine to reach the speed requested, and the speed and distance traveled to reach it. It is interesting to use this information to compare different machines' behavior, the gradual progress of one particular machine, etc. Once we know the injection machine's acceleration curve, we can detect and check any potential wear and tear or changes in injection machine behavior when they begin to occur. The screw acceleration test also helps us to understand (by comparing acceleration graphs) why the molded parts are sometimes not identical in identical machines.

3.7 Pressure Response Test

During the injection molding process, at the end of the mold cavity filling, just as the V/P switchover point is reached, a sudden pressure change occurs as the packing and holding phase begins (Figure 3.11). It goes from the dynamic filling pressure to the pack and hold pressure.

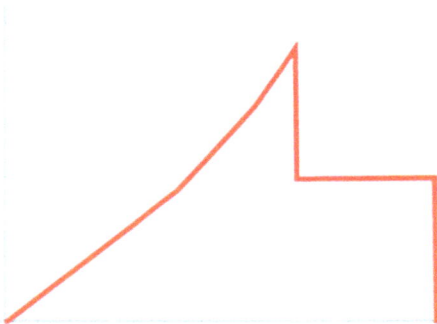

Figure 3.11
Typical plot of injection molding process pressure

This sudden pressure change means that the injection molding machine's control system must promptly regulate very different pressure levels. This is when overtravel occurs (Figure 3.12).

The overtravel is the inertia of pressure, acceleration and deceleration, which occurs upon the change from filling pressure to pack and hold pressure. These inertias mean that the machine, through hydraulic or electrical and electronic control, does not immediately reach the pack and hold pressure set on the machine, and the control system makes some corrections with rapid pressure drops that push the process below the holding pressure setting. This depends on the time it takes for the machine's hardware to communicate the pressure change and the valves' response capacity, so it is also a hydraulic and electronic issue.

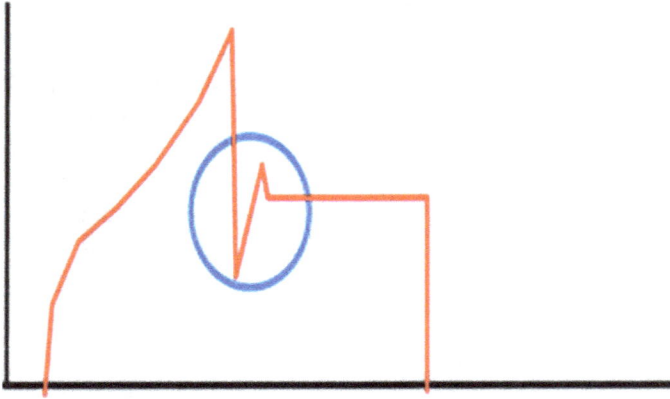

Figure 3.12 Overtravel during the pressure change from dynamic to static pressure

In other cases, to an extent depending on the type of machine, a pressure drop ramp will be executed to reach the packing and holding pressure more smoothly (Figure 3.13). It allows the injection pressure to change and drop gradually over a set period of time after the V/P switchover point has been reached (Figure 3.14). As a result, there is less overtravel but, during the gradual pressure drop, we will be applying not-fully controlled pressure to the cavity rather than suddenly changing the pressure.

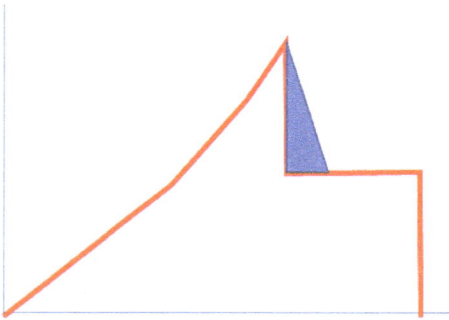

Figure 3.13
Example of ramp pressure overtravel

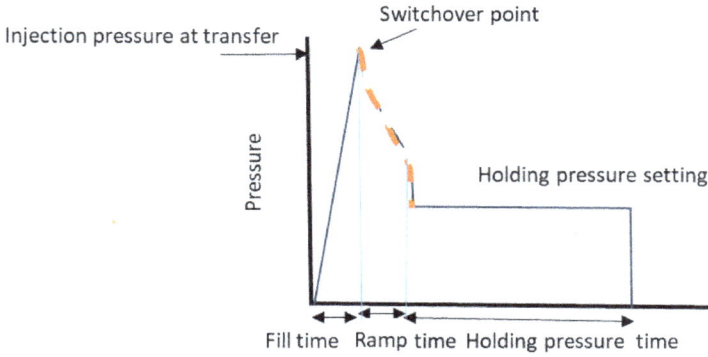

Figure 3.14 Change in ramp pressure

This kind of injection machine behavior indicates the machine's ability to respond to these pressure changes during each injection cycle. The goal for some machine manufacturers is to reduce this uncontrolled overtravel as much as possible.

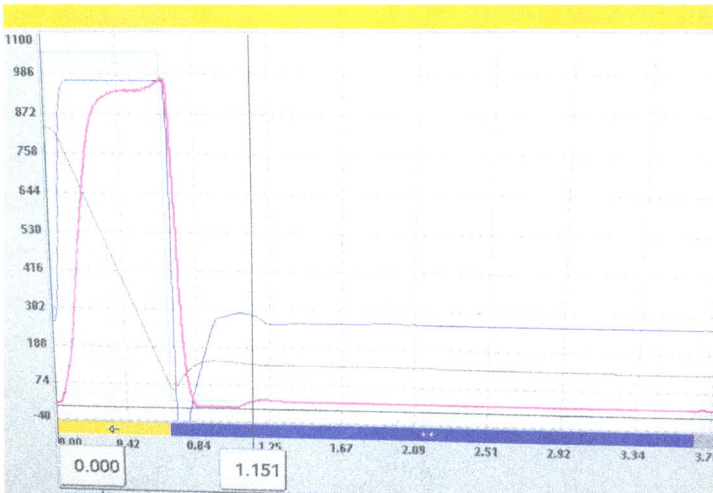

Figure 3.15 Machine real Injection molding graph with overtravel and ramp

In order to be able to control this overtravel and determine the injection machine's ability to withstand pressure changes, we carry out a pressure response test. For this test, we need the pressure and time values at the points P1 and T1, P2 and T2 (as indicated on the graphs in Figure 3.16). These are usually available on the machine control graphs.

The lower the overtravel, time and pressure data, and the faster the injection machine reaches the holding pressure setting, the better is the machine's pressure response.

Pressure Response Test Data and Calculation

Calculation of the machine pressure response level requires the following data, which can usually be found on the injection pressure graph in the machine control system.

Data required:

- P1: pressure peak at the V/P switchover point (sometimes it is not the maximum injection pressure reached)

- T1: mold filling time during the dynamic injection stroke

- P2: pack and hold pressure once the set pressure has been reached

- T2: real injection time elapsed when the holding pressure has been reached

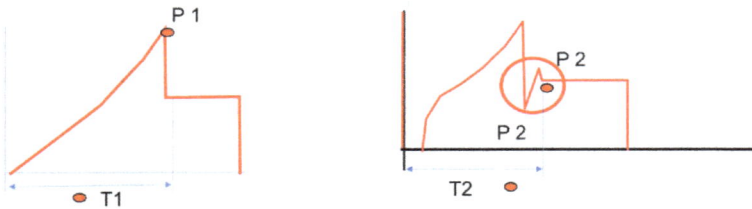

Figure 3.16 Pressure and time data required for the machine pressure response calculation

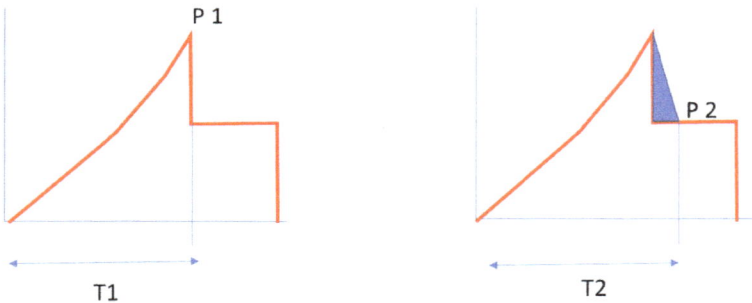

Figure 3.17 Pressure and time data required for the machine pressure response calculation in the case of ramp overtravel

The machine pressure response is then calculated as follows:

A = T2 – T1 pressure time difference

B = (P1 – P2) ÷ 1000 (1000 for specific pressure; 100 for hydraulic pressure)

Machine pressure response = A ÷ B

Maximum acceptable value is 0.2–0.3 (seconds per 100 bar hydraulic pressure or 1000 bar specific pressure)

> Example of machine pressure response calculation:
>
> Pressure response = A ÷ B | A = T2 – T1 | B = (P1 – P2) ÷ 1000
>
> P1 = 1000 bar
>
> P2 = 500 bar
>
> T1 = 2.5 s
>
> T2 = 2.7 s
>
> A = 0.2 s (response time)
>
> B = (1000 – 500) ÷ 1000 = 0.5
>
> Pressure response = 0.4
>
> For a pressure response value of maximum 0.3 s, the machine response time should be 0.15 s; in other words, a pressure response of 0.3 s per 1000 bar of pressure drop difference.

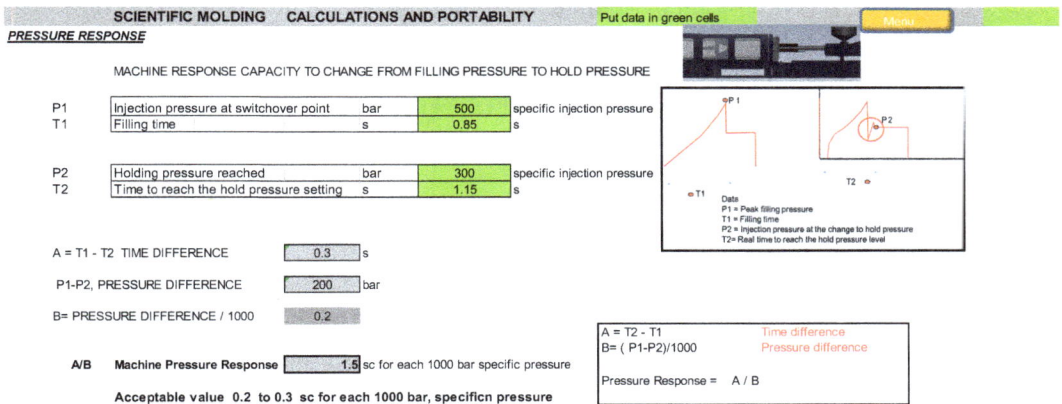

Figure 3.18 A sample spreadsheet for the machine pressure response test

3.8 Other Tests

There are several additional tests that may be carried out to check an injection molding machine's condition.

- Screw position control precision test
- The precision of the machine in controlling the screw position is essential – this precision will directly affect the precision of the metering and the precision at the point of V/P switchover to holding pressure. These two parameters are fundamental to the process repeatability. Check:

- That the screw does not move from the position reached at the metering stroke, during opening, removal, closing, movements
- That the real V/P switchover position is stable and repeatable
- That the metering position reached is stable and repeatable

- Over-injection pressure or inertia test

- This test enables you to analyze the machine repeatability with respect to the position of the V/P switchover point. To carry out this test, the injection molding machine must be configured as follows:
  - Set the holding pressure to zero
  - Set the holding pressure time to zero
  - Select switching by stroke or volume

  In different injection shots, check that the real switchover point at which the machine executes the switch to holding pressure is equal to the set V/P switchover point, screw position or volume. Carry out this test at different speeds (low, medium, and high) to determine the influence of acceleration, deceleration, inertia, hydraulic control, positional control, electronic response, etc. and how they affect the real change in position of V/P switchover to holding pressure.

- Back pressure test

- Test to check that the set back pressure is equal to the real back pressure achieved by the machine. Plot the real back pressure against the set back pressure to control the possible differences.

- Clamping unit control:
  - Check clamping platen parallelism with highly accurate mechanical instruments
  - Check possible platen deflections when applying the clamping force
  - Check clamping force pressure controllers and clamping bar strain sensors
  - These checks and more are usually included in the preventive maintenance of the injection molding machine's clamping system.

- Zero-point screw position test

- Ensure that the zero-point screw position is perfectly under the control of the injection molding machine and check that the screw position is correctly read by the machine. There is nothing worse than thinking there is an adequate cushion available to ensure the appropriate pressure transmission to the cavity and realizing that it is a "fake" cushion that does not really exist as the zero position is incorrect.
  - Manually advance the screw to the maximum forward position
  - Check the machine control data, which should be zero
  - Move the screw to the maximum metering stroke position and check that the machine control data is correct

If the data is not correct, this fundamental position control for the process should be recalibrated and reviewed.

- Safety control check

- Safety is of the utmost importance and must be a priority for everybody, so safety checks must be carried out on safety elements, sensors, position detectors, etc. We cannot accept any safety feature that is not in perfect working order.

3.9 Application Diagram of Injection Molding Machine Reliability and Performance Tests

We propose a comprehensive flow diagram for the application of all these tools or tests for checking the reliability and performance of our machines that allow us to take advantage of the different injection molding machine configurations and thus to obtain data. Although the following sequence is recommended, it is not necessary to follow it exactly in the order given to get the test results. You can follow your own sequence of tests by applying specific tests for the specific analysis of a machine characteristic or its performance.

Application of tests (logical sequence):

Machine Reliability and Performance Tests

- Mold-filling-time repeatability test (explained in detail in Section 3.1)

- Load sensitivity test (explained in detail in Section 3.2)

- Delta P test (explained in detail in Section 3.3)

- Screw tip/check ring valve sealing test (explained in detail in Section 3.4)

- Injection-speed linearity test (explained in detail in Section 3.5)

- Screw acceleration test (explained in detail in Section 3.6)

- Pressure response test (explained in detail in Section 3.7)

- Other possible tests (explained in detail in Section 3.8)

Using all these tools and tests, we will have monitored and controlled the real condition of the injection machines from a precision, precision and performance perspective. These tools enable us not only to anticipate necessary machine maintenance or recalibration interventions, but also to monitor the possible wear and tear on critical machine elements and systems that can cause changes and more variability in the process.

Knowing the Reliability and Performance of Injection Molding Machines: Application of Injection Molding Machine Reliability and Performance Test Tools

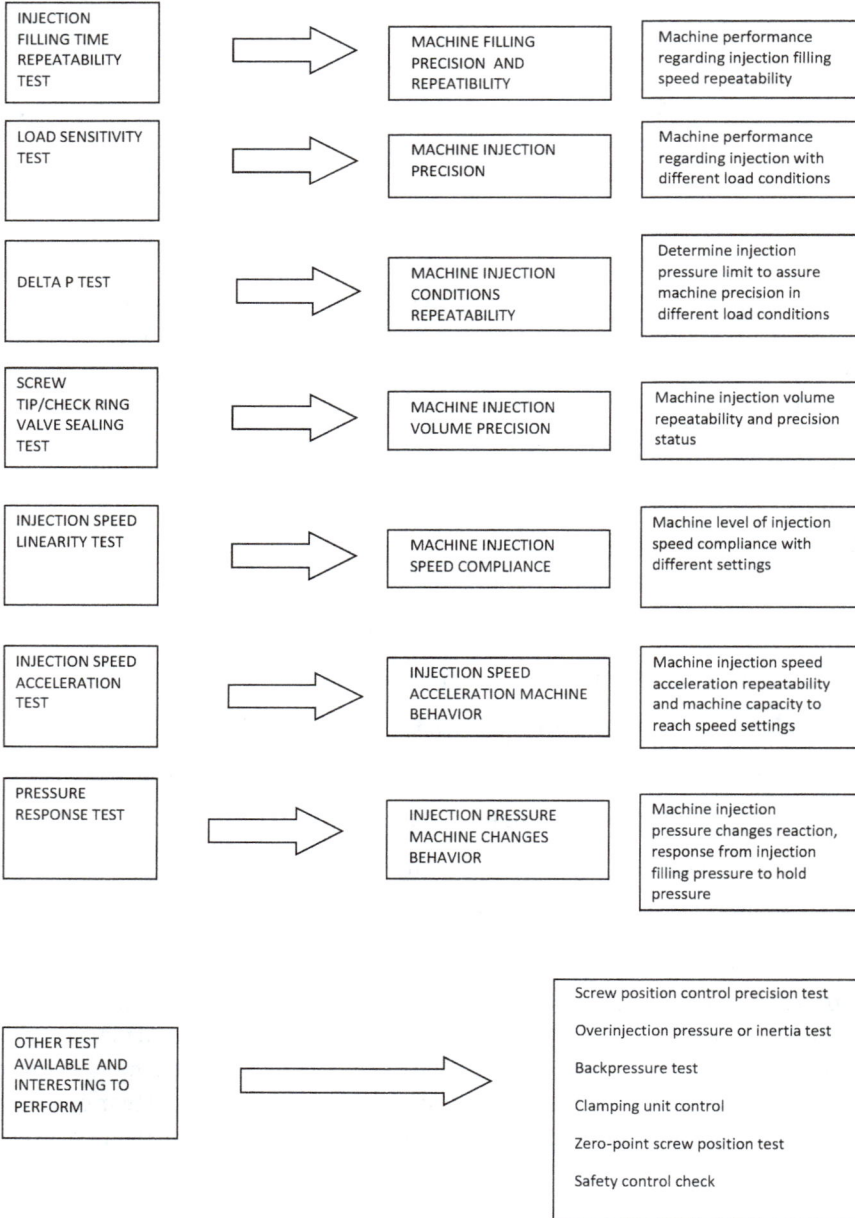

| INJECTION FILLING TIME REPEATABILITY TEST | → | MACHINE FILLING PRECISION AND REPEATIBILITY | Machine performance regarding injection filling speed repeatability |

| LOAD SENSITIVITY TEST | → | MACHINE INJECTION PRECISION | Machine performance regarding injection with different load conditions |

| DELTA P TEST | → | MACHINE INJECTION CONDITIONS REPEATABILITY | Determine injection pressure limit to assure machine precision in different load conditions |

| SCREW TIP/CHECK RING VALVE SEALING TEST | → | MACHINE INJECTION VOLUME PRECISION | Machine injection volume repeatability and precision status |

| INJECTION SPEED LINEARITY TEST | → | MACHINE INJECTION SPEED COMPLIANCE | Machine level of injection speed compliance with different settings |

| INJECTION SPEED ACCELERATION TEST | → | INJECTION SPEED ACCELERATION MACHINE BEHAVIOR | Machine injection speed acceleration repeatability and machine capacity to reach speed settings |

| PRESSURE RESPONSE TEST | → | INJECTION PRESSURE MACHINE CHANGES BEHAVIOR | Machine injection pressure changes reaction, response from injection filling pressure to hold pressure |

| OTHER TEST AVAILABLE AND INTERESTING TO PERFORM | → | Screw position control precision test
Overinjection pressure or inertia test
Backpressure test
Clamping unit control
Zero-point screw position test
Safety control check |

4 Understanding Plastic Materials

Plastics have properties and behaviors that are nothing like those of other substances in the manufacturing industry. Some of these properties and behaviors have to be known by the people involved in transforming or setting up injection molding processes in order to define and control the processes.

In this chapter, we will deal with:

- The classification and properties of plastics according to their molecular structure – amorphous, semi-crystalline, and elastomers – and according to their molecular chain form

- Rheology and viscosity

- Melting and glass transition temperatures

- Viscoelastic behavior

- Creep and relaxation

- Thermodynamic behavior of plastics

- Crystallization

An advanced injection molding technician must know how the polymer that they are going to mold behaves, how its viscosity changes as a function of shear rate, how its specific volume and density change as a function of the injection molding conditions (mainly temperature and applied pressure), what happens to the polymer when packing and holding pressures are applied, and how final shrinkage occurs, for instance. It is also essential to know what T_g is and how it affects plastics and their processing, what crystallization is and its high dependence on the molding conditions applied, etc. A good knowledge of the behavior of the material during the process will allow us to define more repeatable, productive, and robust injection molding processes, as well as to find and determine the causes of certain process deviations that may occur during series production.

4.1 Classification and Properties of Plastics According to Molecular Structure

Thermoplastics are materials that can be softened by the application of heat and hardened by cooling. This softening and solidification process (over a range of temperatures typical of each material) can be repeated several times.

Thermoplastics consist of intertwined polymer chains. In solid form, these chains are unable to move. When heat is applied, this energy causes vibrations in the system, breaking the bonds between the macromolecules, causing them to slide against each other and giving rise to softening and plastic flow. At room temperature, these thermoplastics may be soft or hard and brittle or tough.

Plastics may be classified according to their molecular structure:

- Amorphous thermoplastics

- Semi-crystalline thermoplastics

- Thermoplastic elastomers

4.1.1 Amorphous Thermoplastics

Amorphous thermoplastics have a shapeless and disorganized molecular structure, like a plate of spaghetti (Figure 4.1). The polymer chains are not arranged in any predetermined order but are randomly intertwined. Amorphous thermoplastics contain very long, highly branched molecular chains. Due to their size and shape, these chains cannot be packed compactly and lack structural order. This is why they are called amorphous.

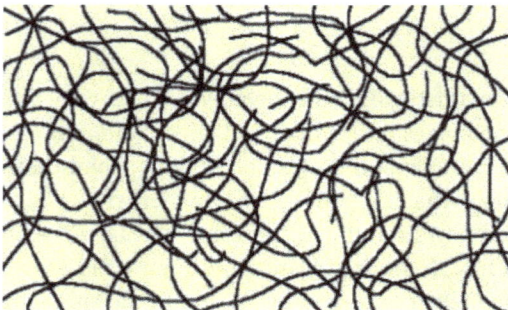

Figure 4.1
Amorphous thermoplastics melt over a fairly wide range of temperatures

Examples of amorphous thermoplastics:

- PS: polystyrene
- PVC-U: rigid polyvinyl chloride
- PVC-P: flexible polyvinyl chloride
- ABS: acrylonitrile butadiene styrene
- SAN: styrene-acrylonitrile
- SB: styrene-butadiene
- PMMA: polymethyl methacrylate
- PC: polycarbonate

Figure 4.2
Molecular structure of an amorphous thermoplastic

Some characteristics of amorphous thermoplastics (see Table 4.1 and Table 4.2):

- Dimensional stability
- Isotropy
- Possibility of transparency
- High viscosity
- More uniform shrinkage
- Tendency to internal stress
- Low post-shrinkage
- Low chemical resistance (stress cracking)

4.1.2 Semi-Crystalline Thermoplastics

Semi-crystalline thermoplastics have an ordered molecular structure, with some molecules aligned with each other to form crystalline regions called crystallites. Semi-crystalline thermoplastics always have an amorphous region and a crystalline region (Figure 4.3).

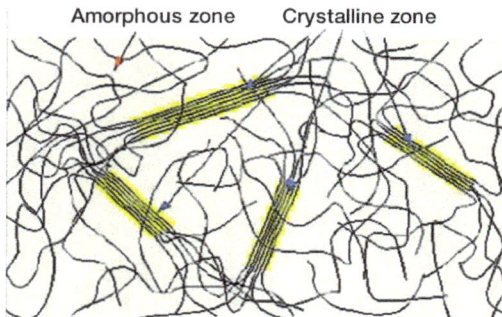

Amorphous zone　　　Crystalline zone

Figure 4.3
Semi-crystalline thermoplastics

The macromolecules in these thermoplastics have few and short side branches. As a result, it is possible that some regions of the molecular chains are sorted and packed against each other. These regions of packed molecular chains are called crystalline regions. Complete crystallization of all regions of the polymer does not occur: there are always disordered or amorphous regions. This is why they are called semi-crystalline polymers.

Examples of semi-crystalline polymers:

- PE-LD: low-density polyethylene
- PE-HD: high-density polyethylene
- PP: polypropylene
- PA: polyamide
- POM: polyoxymethylene (acetal resins)
- PET: polyethylene terephthalate
- PBT: polybutylene terephthalate
- PPS: polyphenylene sulfide
- PEK: polyether ketone
- PTFE: polytetrafluoroethylene

Figure 4.4
Molecular structure of a semi-crystalline thermoplastic

Some characteristics of semi-crystalline plastics (see Table 4.1 and Table 4.2):

- Chemical resistance
- Fatigue resistance
- Reduced presence of internal stresses
- Anisotropy: molded with differential shrinkage
- Greater fluidity

Table 4.1 Generic Differences between Amorphous and Semi-Crystalline Thermoplastics

Amorphous	Semi-Crystalline
Transparent, not all amorphous polymers are transparent	Translucent or opaque
Low molding shrinkage	Extensive molding shrinkage
Less defined melting temperature	Concrete and narrow melting range (3–5 °C)
Moderate mechanical resistance, low fatigue resistance	High mechanical resistance (especially to fatigue and creep)

Table 4.2 Amorphous and Semi-Crystalline Thermoplastics – Generic Properties Comparison

Properties	Crystalline	Amorphous
Specific weight	+	–
Tensile strength	+	–
Tensile modulus	+	–
Ductility, elongation	–	+
Continuous operating temperature	+	–
Fluidity	+	–
Chemical resistance	+	–
Shrinkage	+	–

4.1.3 Elastomers

Elastomers are polymers that are interconnected by a three-dimensional network of cross-links. These cross-links provide elasticity, allowing the material to return to its original shape after being stretched. The cross-links restrict the movement of the polymer chains but still allow some flexibility, giving elastomers their unique elastic properties.

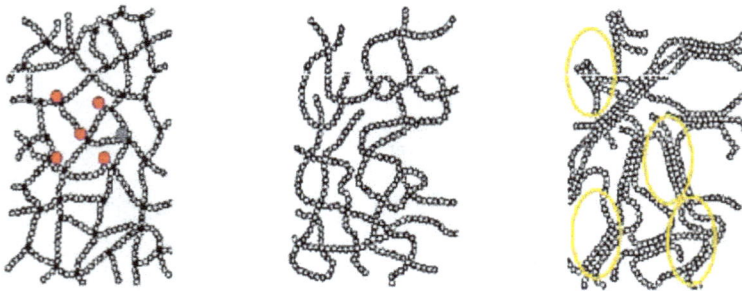

Figure 4.5 Left: thermostable cross-linked molecular network; center: amorphous entangled linear molecules; right: semi-crystalline linear molecular crystals

4.2 Classification of Thermoplastics According to Molecular Chain Form

4.2.1 Homopolymer

Consists of a single monomer. Molecules are composed of identical chemical units. The repeating unit is the same throughout the molecule:

- Linear homopolymer

  A-A-A-A-A-A-A-A-

- Branched homopolymer

  A-A-A-A-A-A-A-A-A-A

       A   A

       A   A

       A

4.2.2 Copolymer

Consists of two or more monomers (i. e. there are two or more different repeating units in its structure):

A-A-A-A-B-B-A-A-A-A-B-B-A-A-A-A-B-B-B-A-A-A-B-B-A-A-

Block copolymer: The monomer sequence is repeated in blocks.

AAAAA-BBBBB-AAAAA-BBBBB-AAAAA-BBBBB-AAAAA-

Random copolymer:

A-B-B-A-A-A-B-B-A-A-B-B-B-B-A-A-B-B-A-A-B-B-A-A-B-B-B-A-A-B-B-A-

Alternating copolymer:

A-B-A-B-A-B-A-B-A-B-A-B-A-B-A-B-A-B-A-B-A-B

The properties of a polymer containing the same number of A and B units will vary with the order of the units in the chain.

4.3 Rheology

Plastics have properties of both an elastic solid and a viscous liquid. They exhibit viscoelastic behavior.

When stress is applied to them, they deform. This deformation can be instantaneous or continuous over time.

Applying stress to a plastic can produce three types of response:

- Elastic deformation

- Viscous flow

- Break

4.3.1 Elastic Deformation

When a material is instantly deformed by the application of stress and the strain is recovered after the stress is removed, the material behaves elastically.

Hooke's law: the stress applied is proportional to the deformation caused.

4.3.2 Viscosity

Viscosity is the resistance of a fluid to motion. When a fluid is put in motion, it will resist such motion. Newtonian fluids have a linear relationship between the force or pressure applied and the velocity of the fluid. Plastics are non-Newtonian fluids that do not exhibit a linear relationship between the force applied and velocity or deformation.

Viscosity is the relationship between stress and strain rate.

> **Newton's law of viscosity**
> If a force is applied to a fluid, the fluid moves at a speed proportional to the force applied.

In Figure 4.6, V1, V2, V3, and V4 represent fluids with different viscosities. The greater the stress (pressure), the greater the deformation of the fluid. This proportional ratio is called the viscosity. In the case of a Newtonian fluid, the ratio increases in proportion to the increase in pressure.

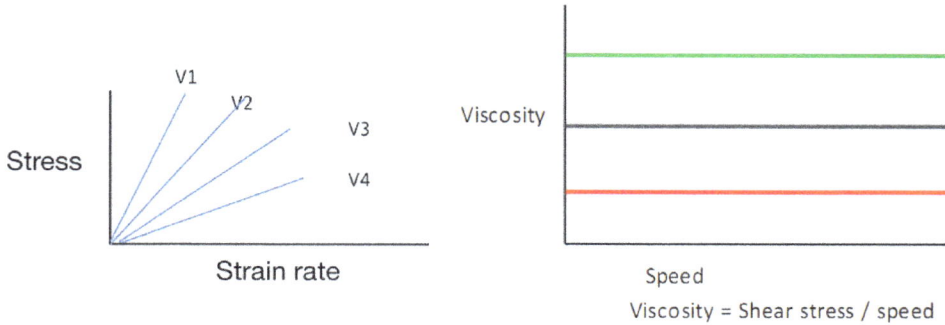

Viscosity = Shear stress / speed

Figure 4.6 Graph showing the viscosity of four different Newtonian fluids exhibiting a proportional stress–strain rate, i. e. there is constant viscosity at different stress and strain levels;

The right graph shows the constant viscosity of different Newtonian fluids ad different shear rate or speed; Newtonian fluids have not influence between speed or shear rate and viscosity, the latter is constant regardless of the shear rate

Plastics are materials that exhibit non-Newtonian behavior.

- They are viscoelastic fluids.
- Viscous behavior violates Newton's law of viscosity.
- The viscosity is not constant.

Plastics are non-Newtonian fluids. When stress is applied to a polymer, the polymer responds with a deformation or strain. The ratio between stress and strain or deformation in polymers is not linear. Therefore, the plastic viscosity depends mainly on the applied stress or shear rate. The higher the shear rate, the lower the viscosity of the same polymer.

In the case of plastic melts in the range of injection molding conditions, an increase in stress can quadruple the deformation or shear rate.

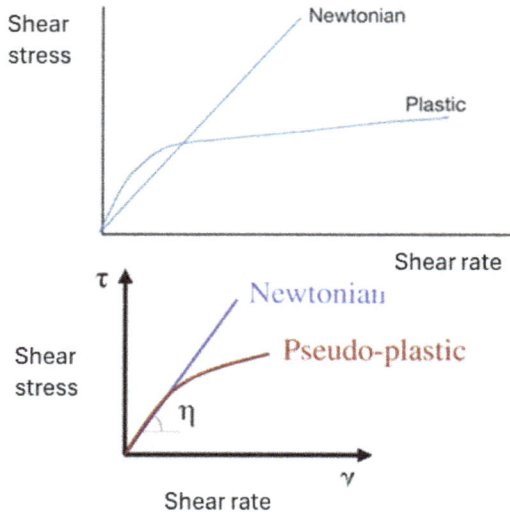

Figure 4.7
Shear stress versus Shear rate
in different behavior, plastics and
Newtonian fluids

In Figure 4.8, we can see that viscosity decreases as the shear rate increases. Figure 4.9 shows that increasing the temperature at low shear rates is more effective at reducing viscosity. However, at high shear rates, the lines of viscosity behavior at different temperatures almost converge and so the temperature has a low influence on viscosity.

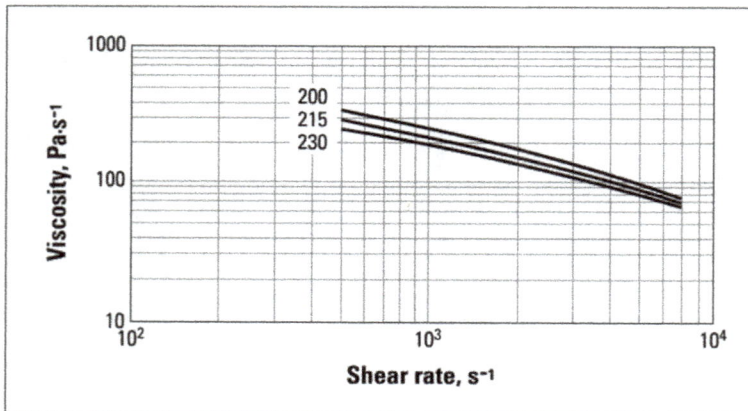

Figure 4.8 An example of viscosity versus shear rate (POM h). Source: DuPont

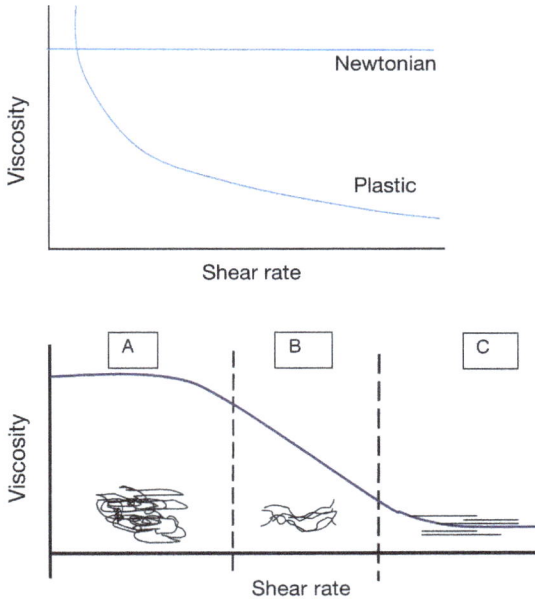

Figure 4.9
The effect of shear on polymer viscosity; A: initial zone, with flat viscosity at low shear rates; B: intermediate zone, with intermediate shear rates and viscosity decreasing with increased shear rate; C: end zone, with high shear rates and flat viscosity; Melt polymers exhibit Newtonian behavior here

Generic shear rate applied to different processes:

- Compression molding: $1–10\ s^{-1}$
- Calendering: $10–100\ s^{-1}$
- Extrusion: $100–1000\ s^{-1}$
- Injection: $1000–10,000\ s^{-1}$

> From the above information, it is evident that if a polymer is processed at a constant flow rate and pressure decrease, the viscosity will be constant, the polymer will flow with the same properties, and it will produce more constant parts dimensions and parts properties.

4.4 Glass Transition Temperature (T$_g$)

The glass transition temperature (T$_g$) is the temperature below which molecular movement is severely restricted and limited. The material is glassy, brittle, and fragile. The movement is restricted and limited to the vibrations of the links and, in some cases, to the rotation of small carbon atoms (no more than 4 or 5). Above the T$_g$, molecular motions are allowed and the material becomes rubbery. Large segments of the chain (more than 100 atoms) are mobile, and one can slide over another (creep). Table 4.3 shows the glass transition temperature of some polymers.

Table 4.3 Glass Transition Temperatures of Some Example Polymers

Polymer	T$_g$ [°C]
PA	20–30
PA 66	57
PP	0 to −20
PC	150
PE	−125
PS	100
PMMA	105
POM	−80
PET	70

4.5 Melting Temperature (T$_m$)

The melting temperature is the temperature at which the crystals formed in semi-crystalline plastics fuse or melt; see Figure 4.10 and Figure 4.11.

CRYSTALLINITY DEGREE AND MELTING TEMPERATURES
OF SOME SEMI-CRYSTALLINE THERMOPLASTICS

MATERIAL	% CRYSTALLINITY	MELTING RANGE (°C)
PELD	40 - 55	105 - 115
PEHD	60 - 80	125 - 140
PP	60 - 70	158 - 168
PIB	50	120 - 130
PA's	< 60	
PA6		215 - 225
PA66		250 - 265
PA610		210 - 225
PA11		180 - 190
PA12		175 - 185
PA46		295
POM (h)	75	175
POM ©	70	165 - 168
PET	30 - 40	255 - 258
PBT	40 - 50	220 - 225
PPS	40	280 - 288
PEEK	40	334
PTFE	53 - 70	327

Figure 4.10
Crystallinity (%) and melting temperatures of some semi-crystalline polymers

SEMI-CRISTALLINE
POLYMER

Degradation temperature

Injection
Extrusion

Rubbery

TEMP °C

TM

Tg

Below Tg: fragile and brittle behavior.

Above Tg the polymer has a tough and elastic
behavior, is a fluid rubbery and has crystals , the
amorphous phase have some moving molecules
,but not the crystals wich remain unmelted .
At the melting themperature Tm, the crystals melt
and the polymer reach a single phase, all the
molecules of the polymer melt.

MOLECULAR WEIGHT

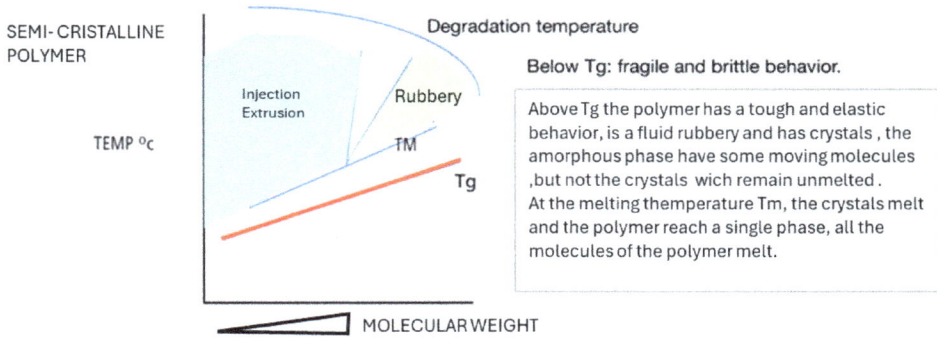

Figure 4.11 Graph of temperature versus molecular weight of semi-crystalline polymers

When an amorphous polymer is exposed to a temperature increase, its chain segments gain mobility. Depending on its molecular weight, the polymer will assume one
or two states (Figure 4.12). Upon reaching the glass transition temperature, the polymer will go from the glassy state to the viscous liquid state (depending on its molecular weight, this could happen before a rubbery state is reached). The higher the molecular weight, the higher the glass transition temperature. The behavior after phase
change T_g may vary from one polymer to another depending on its molecular weight.

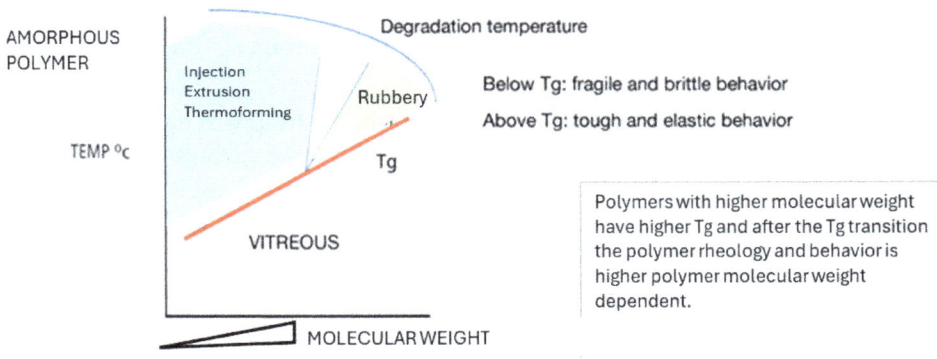

AMORPHOUS
POLYMER

Degradation temperature

Injection
Extrusion
Thermoforming

Rubbery

TEMP °C

Tg

VITREOUS

Below Tg: fragile and brittle behavior

Above Tg: tough and elastic behavior

Polymers with higher molecular weight
have higher Tg and after the Tg transition
the polymer rheology and behavior is
higher polymer molecular weight
dependent.

MOLECULAR WEIGHT

Figure 4.12 Graph of temperature vs. molecular weight of amorphous polymers

4.6 Thermoplastics Behavior

At room temperature (RT), **amorphous plastics** are tough. The molecular bonds hold
the structure together and the molecules can barely move. As the temperature is gradually increased, the macromolecules begin to move and the mechanical tensile
strength decreases, making the material more elastic and tough (Figure 4.13).

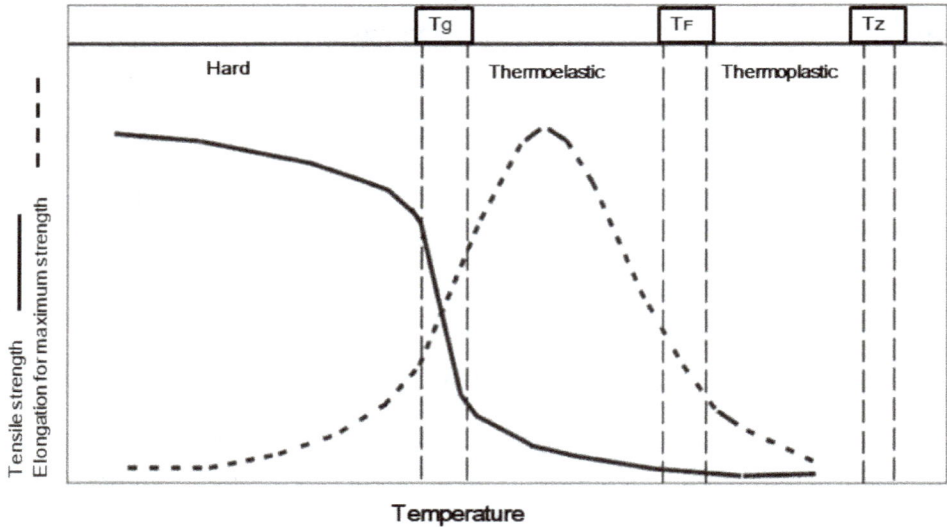

Figure 4.13 Graph showing the behavior of amorphous thermoplastics at different temperatures; $T_G = T_g$, $T_F = T_m$

When the glass transition temperature (T_g) is reached, the intermolecular forces that held the structure together become weak and the macromolecules can slide against each other with relatively little application of external force. Mechanical properties reduce, so elasticity increases abruptly. The material goes from a rigid glassy state to a rubbery state, which is elastic like rubber.

If we increase the temperature, the intermolecular forces disappear, and the rubbery material goes into a molten state (T_m). If the temperature is increased further, the atomic bonds break. Covalent bonds are destroyed, resulting in decomposition and degradation of the material. Part of the molecular weight is lost and the properties of the material are reduced or totally lost depending on the level of degradation reached.

Semi-crystalline thermoplastics, unlike amorphous materials, have two types of structural regions or zones: an amorphous region, where the molecules are spaced apart, and a crystalline region, where the molecules are packed in the form of crystallites. In these crystalline regions, the intermolecular forces are considerably stronger than in the amorphous regions. These crystalline regions reach the thermoplastic range when they are melted on reaching the melting temperature (T_m) (Figure 4.14).

Below the glass transition temperature, semi-crystalline materials are frozen in the amorphous and crystalline regions. Molecular motion is not possible, and the material is hard, fragile, and brittle.

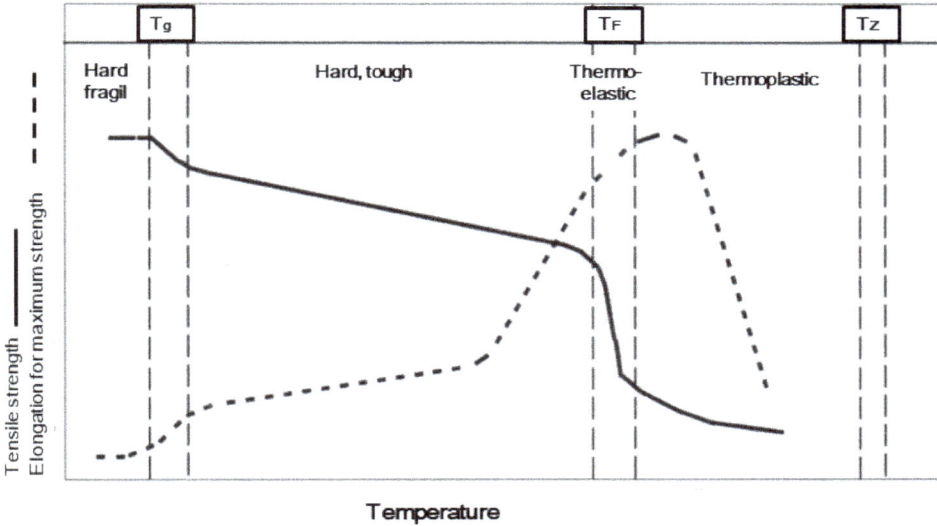

Figure 4.14 Graph showing the behavior of semi-crystalline thermoplastic at different temperatures; $T_G = T_g$, $T_K = T_m$

When the glass transition temperature is exceeded, the first molecules to start moving are those in the amorphous regions due to the intermolecular forces being less intense than in the crystalline regions. Above the semi-crystalline glass transition temperature, materials have toughness properties.

For the most common crystalline plastics, the glass transition temperature (T_g) is below room temperature. As the temperature continues to rise, the molecular chains in the amorphous regions become increasingly mobile, and in the crystalline regions, the molecules slowly begin to vibrate. When the melting temperature (T_m) is reached, the molecules in the crystalline regions are released from the intermolecular forces and the crystallite melting occurs. The molecules of the crystalline regions slide against each other, and all the plastic begins to melt.

As the temperature continues to rise, atomic bonds break. Covalent bonds are broken, resulting in the decomposition and degradation of the material. The molecular weight decreases and the material properties (T_z) will decrease and eventually reduce to zero.

4.7 Behavior under Load

Plastics exhibit different behavior under applied load, the extent of which depends on the load and its duration.

4.7.1 Maxwell–Voigt Model

If a plastic is subjected to a stress or load that does not exceed its strength, it will stretch. If we remove the stress or load, the material will return to its original length. If the stress persists over time, the removal of the stress or load will only restore a fraction of the deformation suffered, and the remainder will be permanently deformed. This effect is called creep or cold flow.

Creep is due to the intricate composition of plastics. Intermolecular forces hold the structure together. When a load is applied, the entanglement is stretched. If the stress ceases without exceeding a certain level, the shape of the tangle will recover. If the load remains, the macromolecules slide against each other and the elongation is irreversible even when the load is removed. See Figure 4.15 and Figure 4.16.

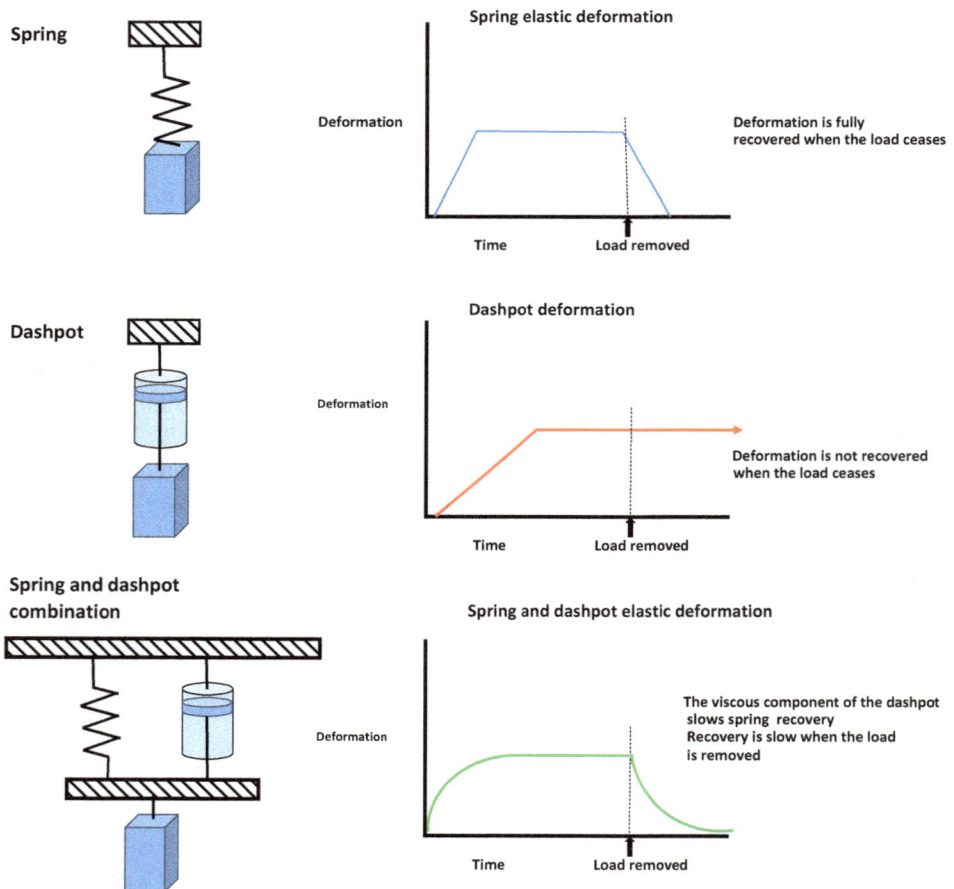

Spring

Spring elastic deformation

Deformation

Time   Load removed

Deformation is fully recovered when the load ceases

Dashpot

Dashpot deformation

Deformation

Time   Load removed

Deformation is not recovered when the load ceases

Spring and dashpot combination

Spring and dashpot elastic deformation

Deformation

The viscous component of the dashpot slows spring recovery
Recovery is slow when the load is removed

Time   Load removed

Figure 4.15 Graphs showing spring and dashpot deformation under load

Maxwell - Voigt Viscoelastic Model

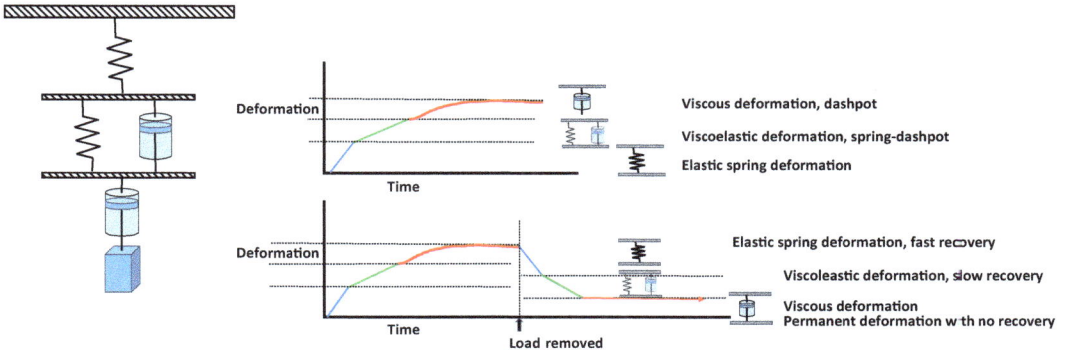

Figure 4.16 Spring and dashpot deformation in plastics

4.7.2 Creep and Relaxation

In Figure 4.17, we can see the tensile creep line at 5.5 MPa stress. Initially, the modulus is tested at 20 MPa. A load of 5.5 MPa produces a deformation of 0.275%. About 10 hours later, the modulus is 50% of the initial value (10 MPa). The deformation at that stage is double the initial value (0.55%).

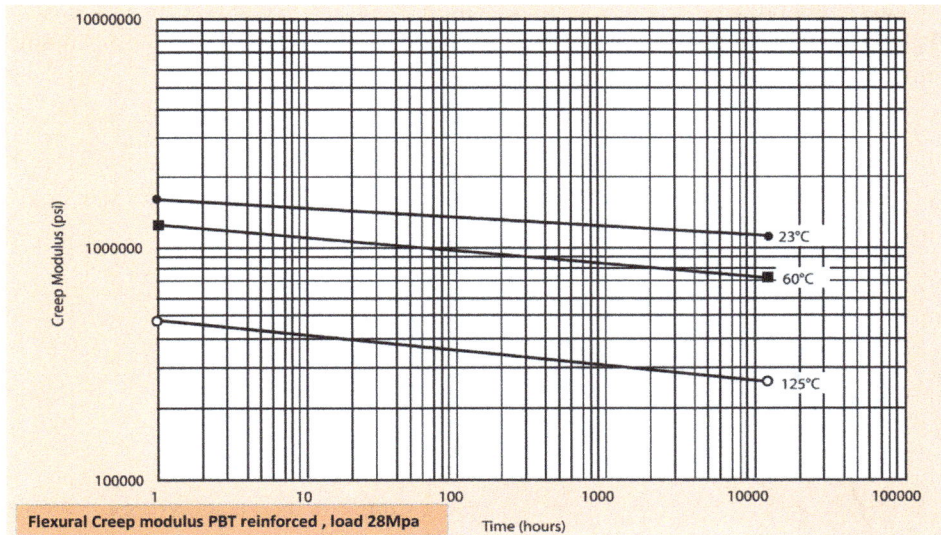

Figure 4.17 Creep graph: an example of the creep modulus of a thermoplastic TPC ET (ASTM 2990) at 23 °C. Source: DuPont

When it comes to designing applications from aspects of stress and time, the creep modulus is the slope of the secant between the origin 0 and the 0t point. Figure 4.18 shows the change in modulus (creep) between 0 and 0t. As the stress is maintained, the strain or deformation increases and the modulus decreases.

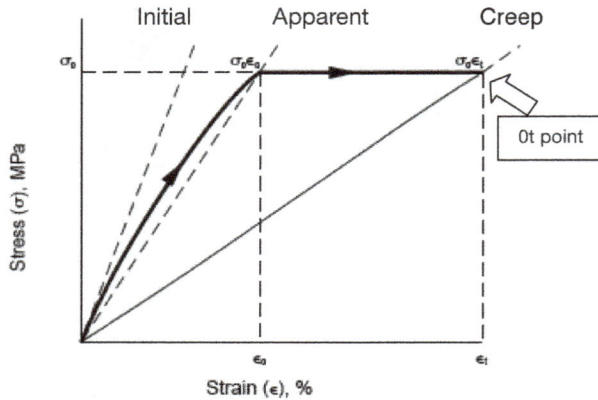

Figure 4.18 An example of a stress–strain graph; As the stress is maintained over time, the strain increases due to creep. Source: DuPont

When designing applications from aspects of relaxation and time (e. g. clippings), the relaxation modulus is the slope of the secant between the origin and the 0t point. In Figure 4.19, we can see how less stress is required while the deformation remains constant due to the relaxation of the polymer.

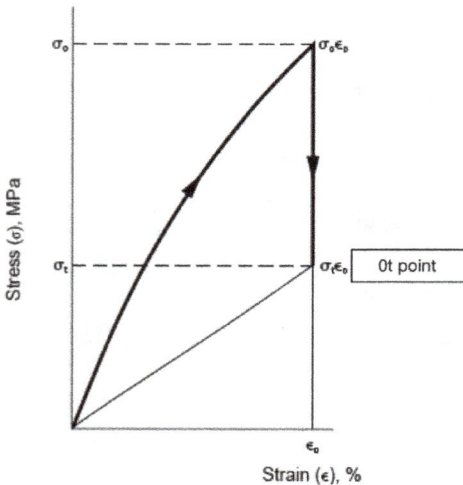

Figure 4.19
An example of a relaxation graph; When the stress is maintained, the stress required decreases due to the relaxation effect. Source: DuPont

4.8 Thermodynamic Behavior of Plastics: PVT Graphs

4.8.1 Thermodynamics Definitions

Specific volume: volume occupied by a given mass (unit: cubic centimeter/gram)

Density: mass of a given volume (unit: gram/cubic centimeter)

Therefore, specific volume is the inverse of density.

4.8.2 PVT Graphs

These graphs illustrate the behavior of polymers in different situations depending on the following variables:

- Pressure
- Specific volume
- Temperature

A PVT (pressure, specific volume, temperature) graph shows the strong relationship that exists between a polymer's specific volume, pressure, and temperature (Figure 4.20 and Figure 4.21). It allows us to represent the injection process and get an idea of the effects of pressure and temperature on the polymer. In other words, we can understand the behavior of the polymer under certain conditions.

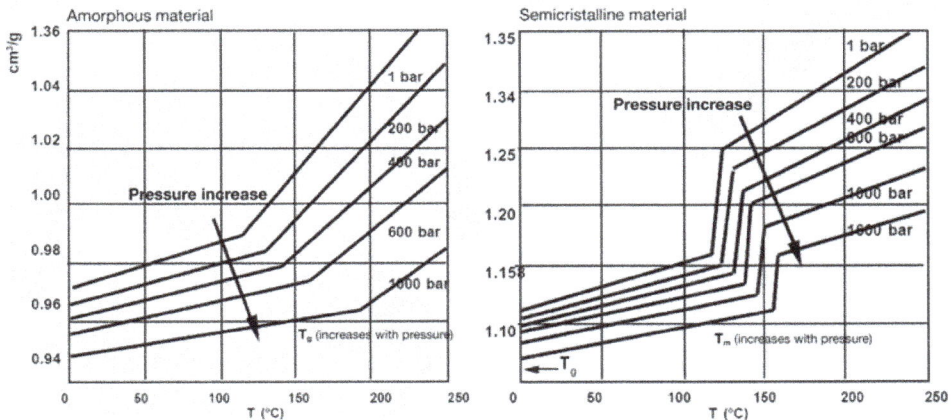

Figure 4.20 Amorphous and semi-crystalline material PVT graphs. Source: Ascamm

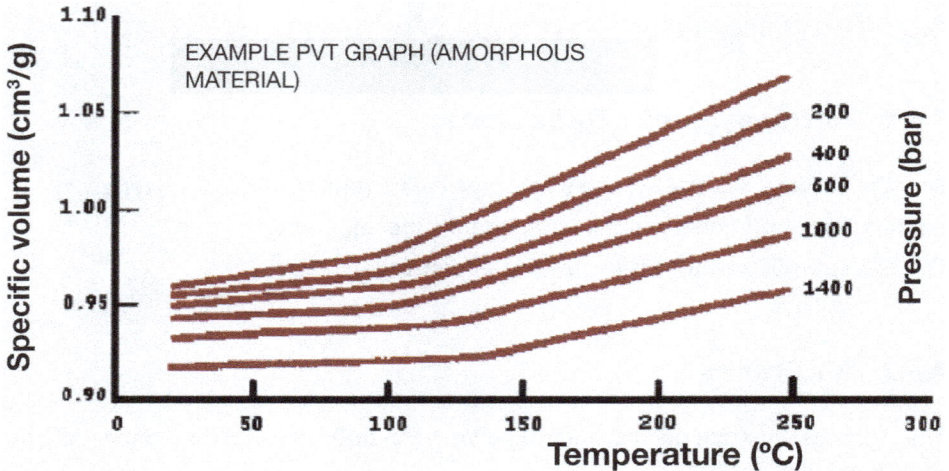

Figure 4.21 Amorphous PVT graph

4.8.2.1 Metering Phase: Plasticization, Melting

When the material is injection molded, it melts as the temperature rises, changing from a solid state (at room temperature) to a molten state (at the processing temperature). During this process, the specific volume increases and the density decreases.

In Figure 4.22, a and b represent:

- a: the specific volume at room temperature

- b: the specific volume at the processing temperature

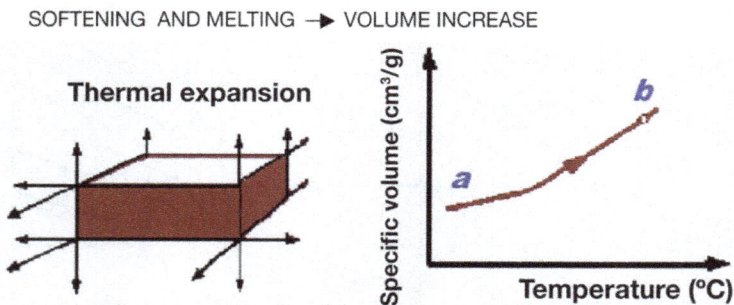

Figure 4.22 The specific volume increases during the melting phase

4.8.2.2 Injection Phase: Filling the Mold

In this phase, the pressure applied to the material is increased. This increase allows the polymer to flow through distribution channels, gates, cavities, etc., overcoming the pressure decreases in the nozzle.

In Figure 4.23, a and b represent:

■ a: the specific volume at process temperature after the metering stage

■ b: the specific volume at the end of filling on the switchover position

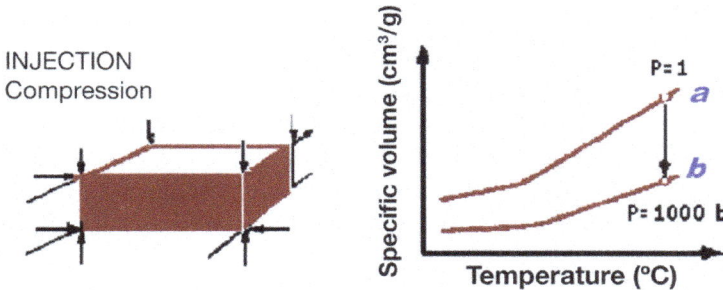

Figure 4.23 During filling, the applied pressure compresses the molten material

As the material exits the plasticizing chamber, it is subjected to injection pressure (compression). During this process, the specific volume decreases and the density increases. Sometimes, the temperature also increases due to the heat generated by friction and shear.

4.8.2.3 Holding Pressure Phase

In Figure 4.24, points 1, 2, and 3 represent:

■ 1: a temperature decrease due to cooling (shrinkage)

■ 2: a pressure decrease (expansion)

■ 3: the best result (constant volume)

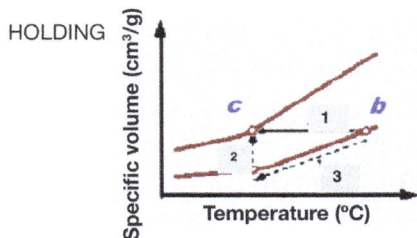

Figure 4.24
Temperature, volume, and pressure effects during the holding pressure phase

When filling is complete, the volume of material in the cavity is about 95–98% of the cavity volume. At this point, the pressure inside the mold is balanced (pressurization phase). If no further compression pressure is applied (holding phase), the pressure in the cavity would suddenly decrease, accompanied by backflow and high shrinkage of the material.

The holding pressure moderates the falling cavity pressure and reaches the atmospheric pressure curve at the required shrinkage level. The ideal process would be to cool the material at a constant specific volume and a steadily decreasing pressure (3), which is equivalent to atmospheric pressure.

4.8.2.4 Cooling Phase

At the end of the previous holding pressure phase, cooling is initiated as the flow rate decreases. The shear stress no longer provides internal heat because there is no large flow movement, and the cooling system removes heat from the mold. After the holding phase, the part continues to cool and will continue to cool as it is removed from the mold until its temperature is in equilibrium with its environment. Cooling is accomplished with decreasing pressure (in the mold) and constant atmospheric pressure (out of the mold).

4.8.3 Main Points of the PVT Graph

See Figure 4.25.

- 1: Start of filling: the specific volume corresponds to the temperature reached and the back pressure applied.

- 1–2: Cavity filling: the pressure increases, and the temperature may also increase.

- 2: Specific volume at the end of filling: pressurization of the cavity begins.

- 2–3: Cavity pressurization phase until switchover point.

- 3–4: Switchover point: a small decrease in pressure may occur due to backflow.

- 4: Compression phase begins.

- 4–5: Pressure is reduced by increasing the cold layer as the part cools. The gate begins to close.

- 5: Gate is sealed. It is not possible to apply pressure in the cavity.

- 5–6: Cooling causes a pressure decrease. The final shrinkage will be due to the reduction in the specific volume during cooling of the material in the mold. This reduction in the specific volume should be partly compensated during compression by adding more molecules into the mold.

- 6: Point of atmospheric pressure. Dimensions are affected by mold shrinkage.

- 6–7: Isobaric cooling (i. e. at constant pressure).

- 7: Opening of the mold and ejection of the part.

- 7–8: Isobaric cooling from the mold. Post-molding shrinkage starts.

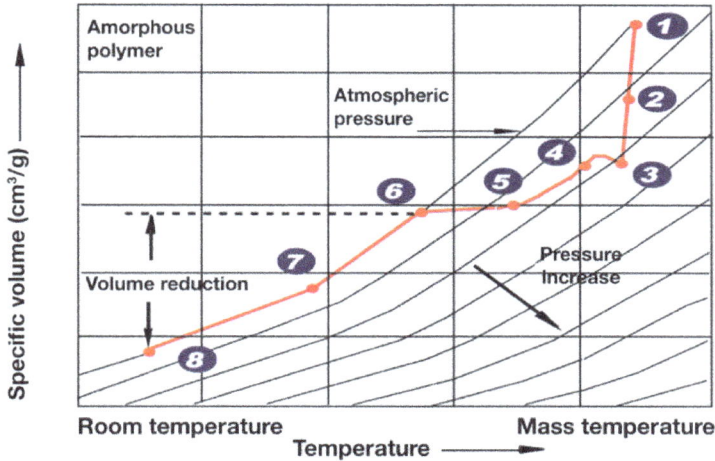

Figure 4.25 PVT graph

4.8.4 Main Effects and Properties of Each Injection Molding Phase on the Material

- Metering
  - Transition from solid to liquid (melting)
  - ↑ Specific volume
  - ↓ Density
  - ↑ Temperature change from room temperature to melting temperature
- Filling
  - High flow volume
  - ↑ Pressure
  - ↑ Temperature-dependent shear stress
  - ↓ Specific volume
  - ↑ Density
- Holding phase
  - Every corner of the part has been filled.
- Holding pressure phase
  - At the end of filling, the pressure inside the mold is balanced.
  - If no pressure is applied, the internal pressure will decrease due to shrinkage.
  - The holding pressure phase should moderate the pressure decrease and end on the atmospheric pressure curve.

- Effects without holding pressure phase, pressure, temperature and volume de-crease with no control: see Figure 4.26.

- Cooling
  - During the holding pressure phase, the flow stops. There is no shear to provide heat.
  - The mold cools the material.
  - Phase 1 – Cooling with mold closed.
  - Phase 2 – Cooling from the mold.
  - Cooling starts at the outer surface and moves toward the core.
  - Recommended removal temperature: 10–15 °C below the Vicat temperature.

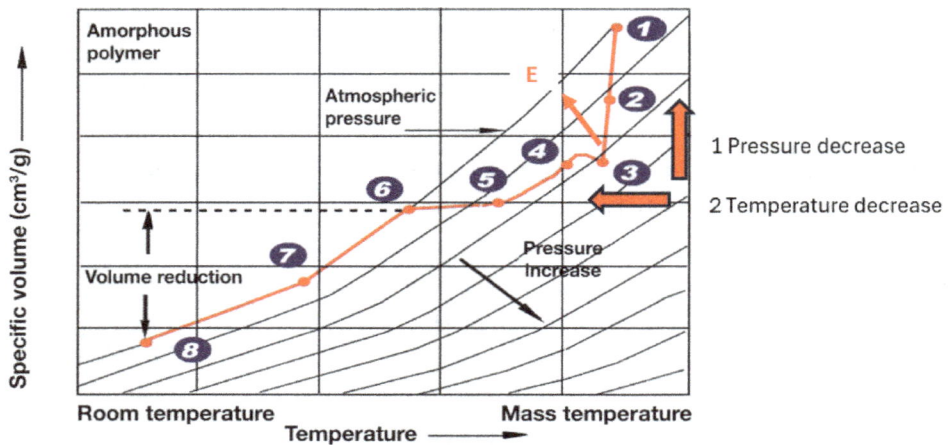

Figure 4.26 Effects without holding pressure phase; pressure, temperature and volume decrease with no control after point 3 in the graph:
1: Temperature decrease
2: Pressure decrease
E: Effective results on pressure and temperature without holding pressure phase
Ideally: keep a volume constant as much as possible between point 3 and point 6 in the graph

In the first PVT graph (top) in Figure 4.27 we can see that the holding pressure time of 5 seconds is insufficient and creates a sharp decrease in pressure. In the second PVT graph (bottom), the compression time is 20 seconds, until the gate is sealed, and the pressure decrease is much more progressive. Cavity pressure is maintained for a more effective time, and shrinkage of the molded part is reduced.

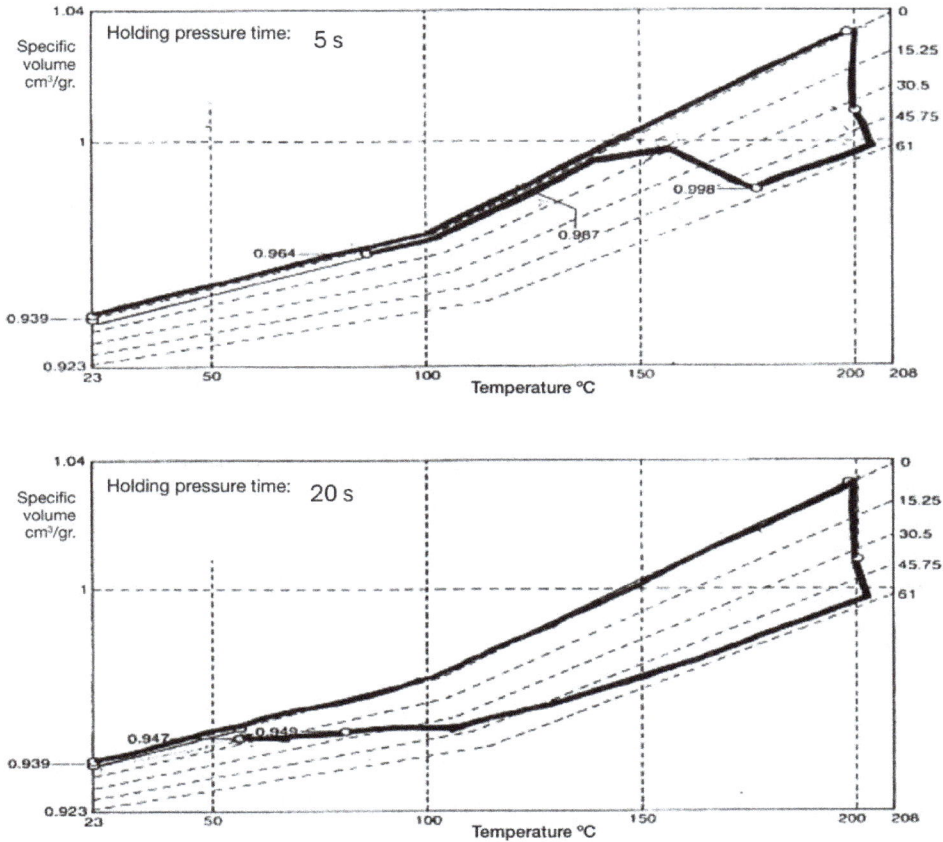

Figure 4.27 PVT graphs: they differ in the holding pressure time. Source: AITIIP Technical Report

4.8.5 Influence of Injection Molding Parameters Reflected in PVT Graphs

Higher Back Pressure

A different constant pressure curve means:

- → Higher density
- → Higher mass dose

Temperature

Higher barrel temperature means higher material temperature and:

- → Higher specific volume
- → Lower mass dose

Filling Pressure

Increased filling pressure and holding time results in:

- → Lower volumetric shrinkage

Injection Speed

Higher friction means:

- → Higher shear
- → Higher temperature
- → Higher specific volume
- → Higher shrinkage

Holding Pressure

- Short = pressure decrease = higher shrinkage
- Long = lower pressure decrease = reduced shrinkage = higher weight

Cooling

Two phases:

- Closed mold
  - Decreasing pressure
- Open mold
  - Constant atmospheric pressure

→ Molding shrinkage

→ Post-molding shrinkage

Figure 4.28 and Figure 4.29 display PVT graphs of a POM h semi-crystalline material. In the two graphs in Figure 4.29, we can see the development of the specific volume over a range of applied temperatures. We can also see this development of the specific volume with different applied pressures. Since the material is a semi-crystalline polymer, the melting of crystals occurs in the temperature range between 180 °C and 200 °C, producing this effect with a rapid increase in the specific volume in the thermal range.

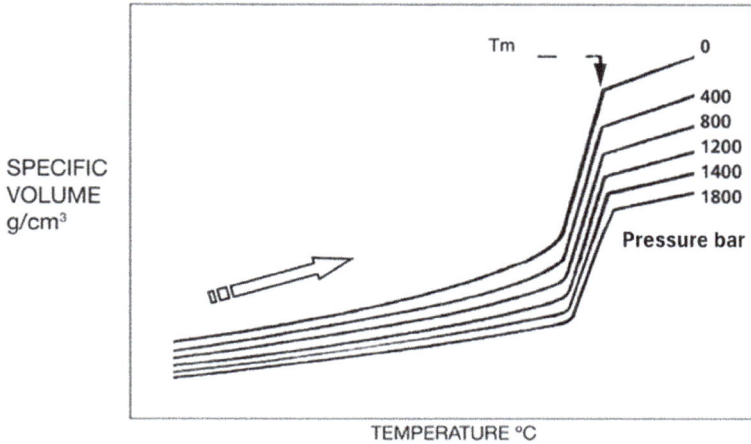

Figure 4.28 PVT graph of a POM h semi-crystalline material. Source: DuPont Technical Information Brochures

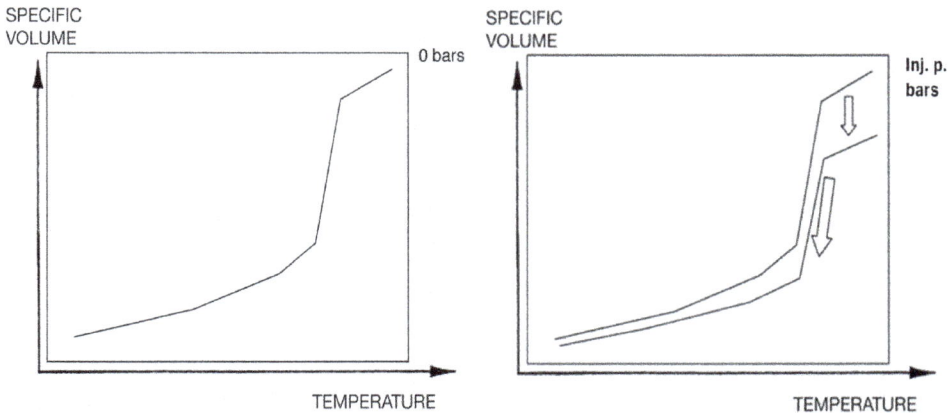

Figure 4.29 Development of the specific volume over a range of applied temperatures.

Left: Metering phase: Material transition from room temperature on the barrel throat to process temperature. During melting: The transition from solid to liquid increases the specific volume of the polymer (16% in the case of POM).

Right: Injection and crystallization: Crystallization/volume shrinkage. Molecular reorganization. The molecular reorganization creates voids that should be filled with more molten material. Crystallization under constant pressure. As we inject, we compress the melt material and crystallization begins

Imagine that the cavity to be filled is a room, the door is the cavity gate, and the corridor is the runner or distribution channel (see Figure 4.30). During the filling process, we fill the room with boxes randomly and chaotically to fill the entire volume, just as we fill the cavity with the molten polymer (Figure 4.31).

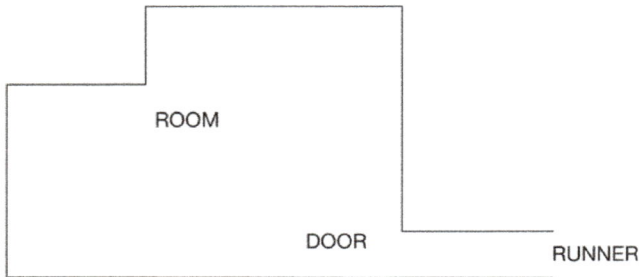

Figure 4.30 Allegory of a semi-crystalline material during the filling, holding and crystallization phases. Source: DuPont

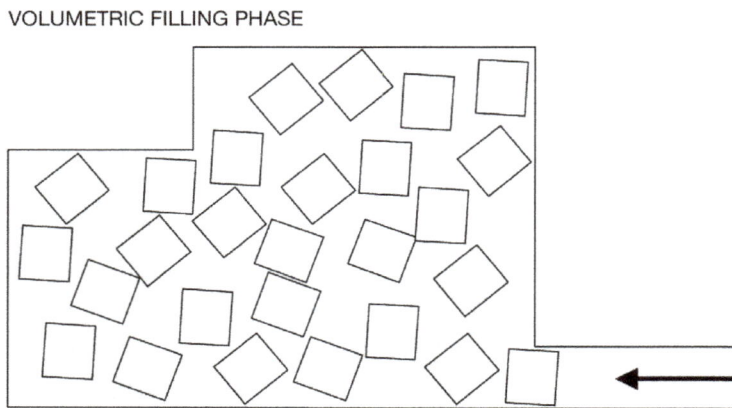

Figure 4.31 Cavity-filling allegory

During cooling, the molecules reorganize themselves and seek thermal and physical equilibrium, so that they occupy less volume than the initial volume. This creates an empty space, a volume that we should compensate or fill in order to obtain higher-quality parts (Figure 4.32).

During the packing and holding phase, it is important not to overcompress (Figure 4.33), with cooling until at least the ejection of the part in solid and hot state, with low cavity pressure over the material.

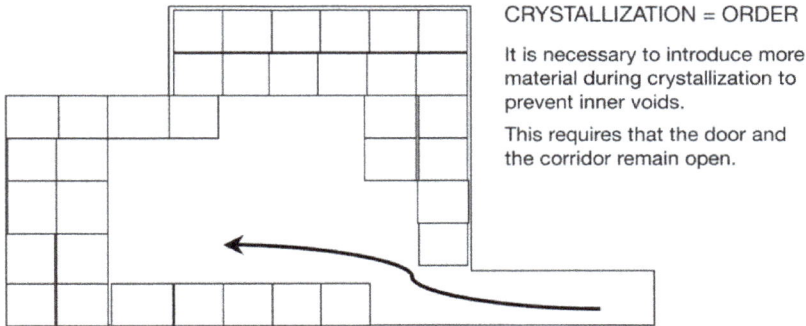

CRYSTALLIZATION = ORDER

It is necessary to introduce more material during crystallization to prevent inner voids.

This requires that the door and the corridor remain open.

Figure 4.32 Cooling, shrinkage, and free-volume allegory

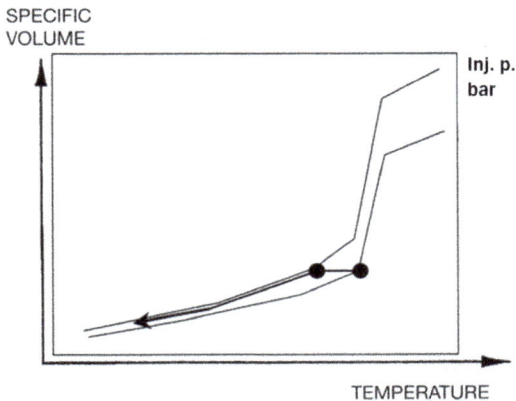

SPECIFIC VOLUME

Inj. p. bar

TEMPERATURE

Figure 4.33 Packing and holding pressure and cooling phases

4.9 Crystallization Phases

The degree of crystallization will vary with the rate at which the polymer cools. If the melt polymer cools quickly, the degree of crystallization will be low, and vice versa if the material cools slowly. As the plastic cools into the cavity, different cooling rates are present on the same section of the part.

First zone: Skin

In this first phase, the material, when in contact with the "cold" mold steel surface cools quickly, forming a solid skin (Figure 4.34 and Figure 4.35). Below this, crystallization occurs in the areas closest to the cold skin. The thickness of this skin therefore gradually increases. This is a freeze layer of low crystallinity.

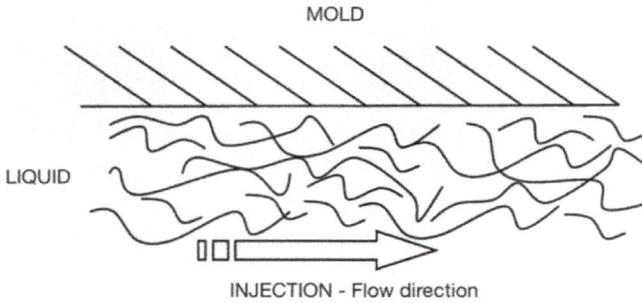

Figure 4.34 Melt polymer flow

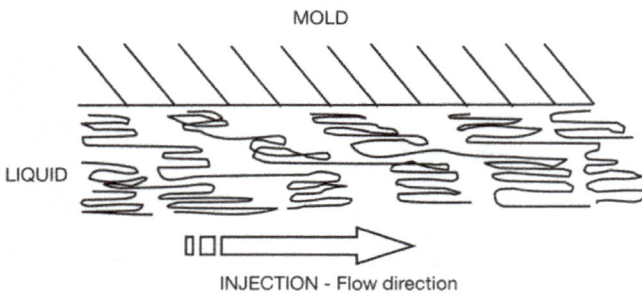

Figure 4.35 Generation of lamellae in the first exterior zone or skin

Intermediate zone

The mold temperature is very important in this intermediate zone (Figure 4.36). The cooling rate affects the formation and size of these lamellae or crystalline structures. Structures obtained in cold molds will be weaker than those obtained in hot molds, which have a slower crystallization rate.

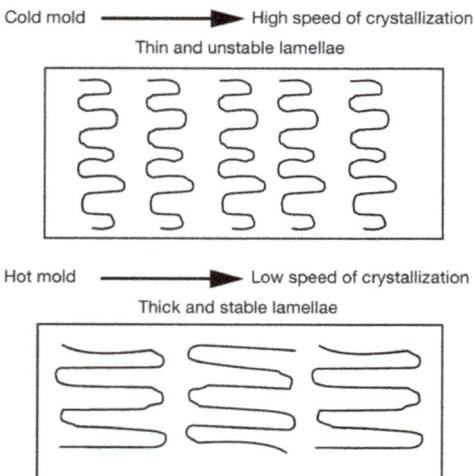

Figure 4.36
Lamellae development in the intermediate zone at different mold temperatures

Central zone

Crystallites are formed due to slower cooling. The material has more time to cool because the interlayer and skin layer isolate the central area of the part thickness.

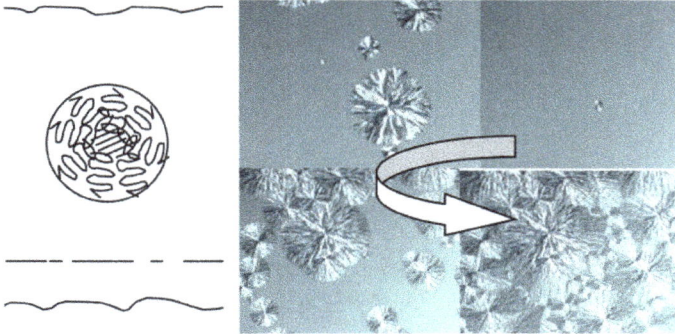

4.9.1 Defects or Errors Caused during the Crystallization Phase

Uncontrolled post-molding shrinkage

Too low a mold temperature will produce thin and unstable lamellae, which will become thicker and more stable as time passes and the temperature stabilizes.

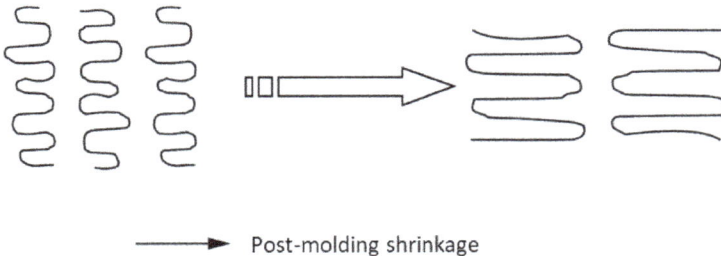

Post-molding shrinkage

Semi-crystalline polymers:

This weak crystallization can lead to post-crystallization or post-shrinkage of the polymer structure, resulting in warping, cracking, or deformation (Figure 4.37). If crystallization has not been achieved at all, when the part is in a high-temperature environment, the crystallization process will recommence and proceed until the most equilibrated state of the material with no internal stress has been reached. This post-crystallization requires more post-shrinkage. If the part cannot shrink because it is being assembled, for instance, the part may break, crack, or deform.

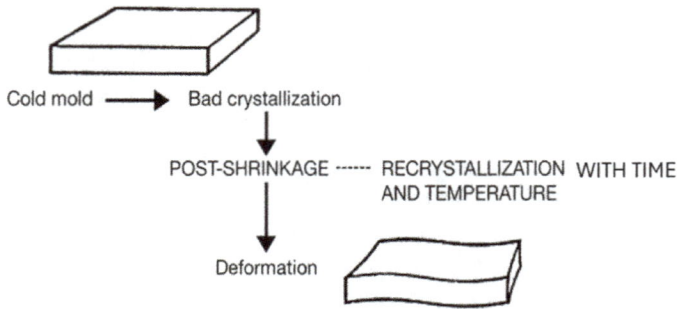

Figure 4.37 Post-crystallization

4.9.2 Degree of Crystallization, an Injection Molding Technician's Responsibility

The degree of crystallization achieved in the part could be very different from the crystallinity property of the plastic material. For example, a POM homopolymer has a potential crystallization of 75% but, to an extent depending on the injection molding process conditions, the part might only achieve 40–50% or even lower.

The ability of a semi-crystalline polymer to crystallize is mainly controlled by the cooling rate or cooling speed of the molten material in the mold cavity. The key parameter is mold temperature. This is the responsibility of the injection molding technician because the mold temperature determines the degree of crystallinity achieved and this degree of crystallinity will greatly influence the properties of the part. Molding with a low mold temperature will give rise to less crystallinity than molding with a hot mold temperature.

The degree of crystallinity of the molded part and the performance of the part with semi-crystalline material are the responsibility of the molder.

5 Required Information for Defining the Process

In order to apply the **scientific injection molding** methodology, we should not only know the tools or studies that it suggests and on which this methodology is based (as mentioned in previous chapters, through the application of these tools we can control and check our machines, the state of precision of our machines, our plastics and their typical behavior in the process). We should also collect a whole set of the basic information required to successfully define the process. This basic general information that we should collect may be grouped into the following groups or families:

- Plastic material
- Part
- Mold
- Summary

We must try to collect the maximum amount of information discussed in this chapter so that we can fully evaluate each decision we make regarding the definition of the process, as well as possible future deviations during production of the part series.

5.1 Plastics Properties: Data Sheet Interpretation

Correctly understanding a plastic technical data sheet will greatly help us get to know what material we are going to mold and its properties, behavior, etc. These are some of the most typical properties of plastic materials – they are used to compare materials as well as to help with selecting the best materials for each application:

- Density
- Melt flow index (MFI)

- Elastic or tensile modulus

- Tensile strength

- Impact strength

- Cold flow (creep)

- Creep modulus

- Softening temperature (Vicat)

- Heat deflection temperature (HDT)

These properties, amongst others, are characterized in the technical data sheet (TDS) published by the plastic manufacturers.

5.1.1 Examples of Data Sheets

Zytel® 70G30HSL NC010 is a 30% glass fiber reinforced, heat stabilized polyamide 66 resin for injection molding.

Property	Test Method	Units	Value	
			DAM	50%RH
Identification				
Resin Identification	ISO 1043		PA66-GF30	
Part Marking Code	ISO 11469		>PA66-GF30<	
Mechanical				
Stress at Break	ISO 527	MPa (kpsi)	195 (28.3)	125 (18.1)
Strain at Break	ISO 527	%	3.4	5
Tensile Modulus	ISO 527	MPa (kpsi)	10000 (1450)	7200 (1045)
Tensile Creep Modulus	ISO 899	MPa (kpsi)		
1h				6800 (990)
1000h				5100 (740)
Notched Charpy Impact Strength	ISO 179/1eA	kJ/m^2		
-30°C (-22°F)			10	10
23°C (73°F)			12	16
Unnotched Charpy Impact Strength	ISO 179/1eU	kJ/m^2		
-30°C (-22°F)			70	73
23°C (73°F)			82	93

Figure 5.1 A mechanical properties data sheet. Source: Dupont

Property	Test Method	Units	Value	
			DAM	50%RH
Thermal				
Deflection Temperature	ISO 75f	°C (°F)		
0.45MPa			261 (502)	
1.80MPa			248 (478)	
Melting Temperature	ISO 11357-1/-3	°C (°F)		
10°C/min			263 (505)	
CLTE, Normal	ISO 11359-1/-2	E-4/C (E-4/F)		
23 - 55°C (73 - 130°F)			1.07 (0.60)	
CLTE, Parallel	ISO 11359-1/-2	E-4/C (E-4/F)		
23 - 55°C (73 - 130°F)			0.22 (0.12)	
Vicat Softening Temperature	ISO 306	°C (°F)		
50N			250 (482)	

Figure 5.2 A thermal properties data sheet

Property	Test Method	Units	DAM	50%RH
Electrical				
Surface Resistivity	IEC 60093	ohm	>1E15	1E13
Relative Permittivity	IEC 60250			
1E2 Hz			4.4	10.8
1E6 Hz			4.1	4.6
Volume Resistivity	IEC 60093	ohm m	>1E15	1E9
Dissipation Factor	IEC 60250	E-4		
1E2 Hz			70	4600
1E6 Hz			150	650
Electric Strength	IEC 60243-1	kV/mm (V/mil)		
1.0mm			38 (964)	32 (812)
CTI	IEC 60112	V	400	
CTI	UL 746A	V		
3.0mm			400	

Figure 5.3 An electrical properties data sheet

5.1.2 Density

The density of a material is the mass per unit volume, usually expressed in g/cm^3. Figure 5.4 shows the density of some polymers.

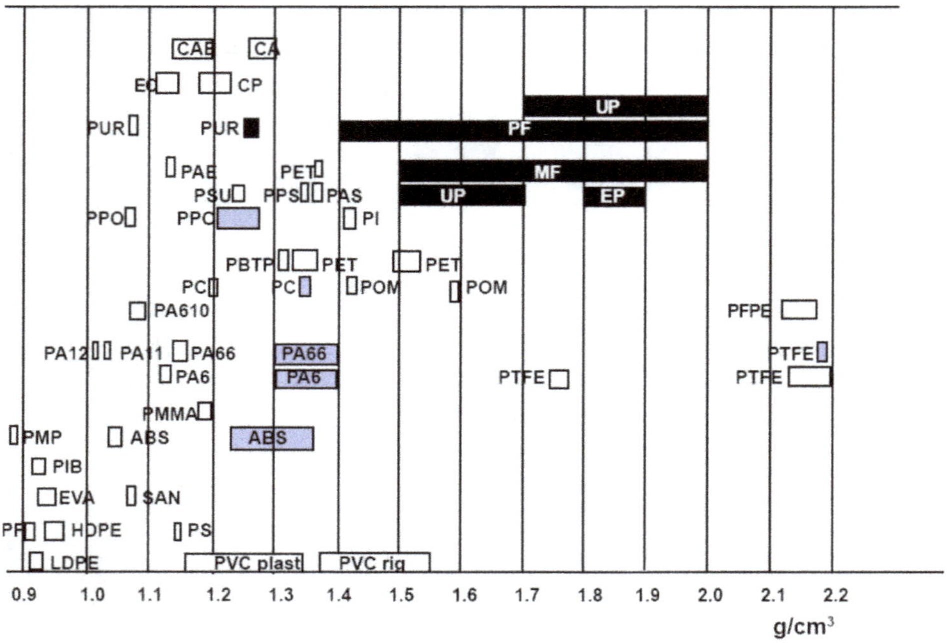

Figure 5.4 The density of various polymers

Keep in mind that density directly affects the cost of the part. The molder buys plastic by unit weight (typically 25 kg bags) and sells parts that have the volume of the cavity in which they were molded. The factor that determines the weight of the part, and therefore its cost, linking these two factors – weight and volume – is density. So, if we manage to mold a part with the right properties using a plastic of lower density, we will reduce the cost of the part since we will be able to mold many more parts per kilogram of raw material.

For example, if we mold PMMA with a density of 1.19 g/cm^3 and the molded part has a volume of 10 cm^3, the weight of the part will be 11.9 grams. With a 25 kg bag, we can produce 2100 parts. If the same part is molded with SMMA polymer (to meet part requirements), the SMMA density is 1.12 g/cm^3 – it could produce 2232 parts with a 25 kg bag, leading to an increase in productivity of more than 6%!

5.1.3 Material Flow Rates

Various material flow rate data are used and published by the polymer manufacturers. All of them attempt to indicate the material viscosity.

5.1.3.1 Melt Volume Index (MVI)

The MVI is the most widely used reference for determining a material's fluidity. It is obtained with the aid of a capillary rheometer (see Figure 5.5). A quantity of pellets (previously dried) is placed in the device and heated to the standard test temperature (depending on the material; see Table 5.1). Once it melts, a standard weight (also depending on the material to be tested; see Table 5.1) is applied so that it flows through a calibrated outlet. A certain amount of material flowing through the rheometer is therefore obtained at a given time. Flow values are expressed in cubic centimeters per 10 minutes ($cm^3/10$ min).

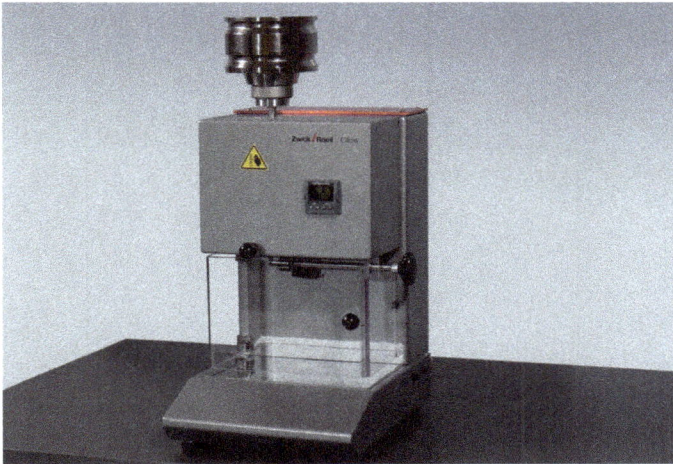

Figure 5.5 Laboratory capillary rheometer. Source: Zwick Roell

5.1.3.2 Melt Flow Index (MFI)

The MFI, expressed in grams per 10 minutes, is obtained in a similar way to the MVI. It is necessary to multiply the value obtained by the volume density of the molten material at test temperature. It can also be obtained by weighing the amount of molten material that flows through the capillary rheometer.

Table 5.1 Standard Conditions for MFI Tests (ASTM D1238 and ASTM D3364)

Condition	Temperature (°C)	Load Weight (kg)	Pressure (kg/cm^2)
A	125	0.325	0.46
B	125	2.160	3.04
C	150	2.160	3.04
D	190	0.325	0.46

Table 5.1 Standard Conditions for MFI Tests (ASTM D1238 and ASTM D3364) *(continued)*

Condition	Temperature (°C)	Load Weight (kg)	Pressure (kg/cm²)
E	190	2.160	3.04
F	190	21.600	30.40
G	200	5.000	7.03
H	230	1.200	1.69
I	230	3.800	5.34
J	265	12.500	17.58
K	275	0.325	0.46
L	230	2.160	3.04
M	190	1.050	1.48
N	190	10.000	14.06
O	300	1.200	1.69
P	190	5.000	7.03
Q	235	1.000	1.41
R	235	2.160	3.04
S	235	5.000	7.03
T	250	2.160	3.04

Table 5.2 Standard MFI Test Conditions for Different Materials (ASTM D1238 and ASTM D3364)

Material	Test Condition
Acetals (POM)	E, M
Acrylics	H, I
Acrylonitrile butadiene styrene (ABS)	G
Cellulose esters (CE)	D, E, F
Nylon (PA)	K, Q, R, S
Polychlorotrifluoroethylene (PCTFE)	J
Polyethylene (PE)	A, B, D, E, F, N
Polyethylene terephthalate (PET)	T

Material	Test Condition
Polycarbonate (PC)	O
Polypropylene (PP)	L
Polystyrene (PS)	G, H, I, P
Vinyl acetal (PVC)	C

Table 5.3 Standard MFI Test Conditions for Different Materials (ISO 1133)

Condition	Temperature (°C)	Load (kg)
A	250	2.16
B	150	2.16
D	190	2.16
E	190	0.325
F	190	10.00
G	190	21.6
H	200	5.00
M	230	2.16
N	230	3.8
S	280	2.16
T	190	5.00
U	220	10.00
W	300	1.2
Z	125	0.325

5.1.3.3 Comparing Materials' Fluidity

Be particularly careful when comparing the fluidity of two or more materials.

The capillary rheometer fluidity test (Figure 5.6) is what we can call a "static" test, meaning that very little shear is applied to the polymer melt during the test. It is evident that when a weight is applied, according to the standard fluidity index test, the stress applied to the plastic is held constant and at a low shear rate.

These test conditions are very different from the injection process conditions, where the injection machine applies increased stress or injection pressure over the melt material and a higher shear rate. When plastics are subjected to the usual pressure and shear rate in an injection molding process, their rheological behavior differs from what can be obtained in this fluidity test. The decrease in viscosity under these molding process conditions will be very different between two polymers that may exhibit similar fluidity in the fluidity test.

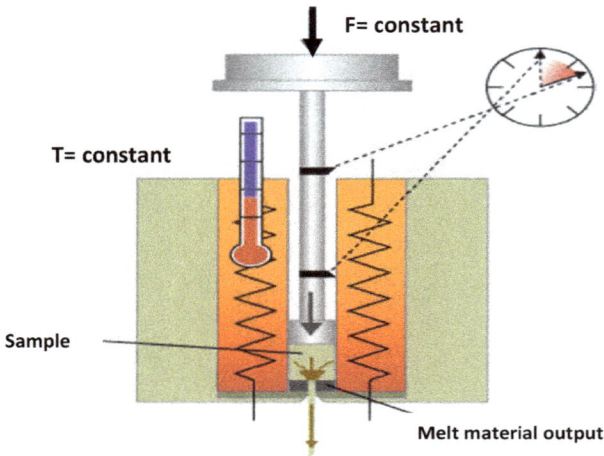

Figure 5.6
Fluidity test diagram

5.1.4 Melting Temperatures

Table 5.4 shows the typical temperatures at which semi-crystalline materials' structures melt. The melting temperature is the temperature that must be reached in order to melt the crystallites and therefore obtain a good homogenization of the melted polymer. Below this temperature, semi-crystalline materials are somewhat rigid, as the crystallites' structure is yet to melt.

Table 5.4 Melting Temperatures of Some Polymers

Material	Temperature (°C)
PS	70–115
PE-HD	125–140
PE-LD	105–115
PMMA	120–160
PP	160–170
PA11	180–190

Material	Temperature (°C)
PA6	215–225
PBT	220
PC	220–230
PA66	250–260
PET	250–260

5.1.5 Mechanical Properties

The stiffness of the material and its resistance to breakage when subjected to tensile stress, its elasticity, and its impact strength, amongst others, are vital properties that we should be aware of and understand.

5.1.5.1 Tensile Stress and Mechanical Resistance

This test determines the stiffness of the material and its resistance to breakage when subjected to tensile stress (Figure 5.7). It can be conducted at different temperatures and stretch rates. In this way, different characteristics of the polymer can be obtained.

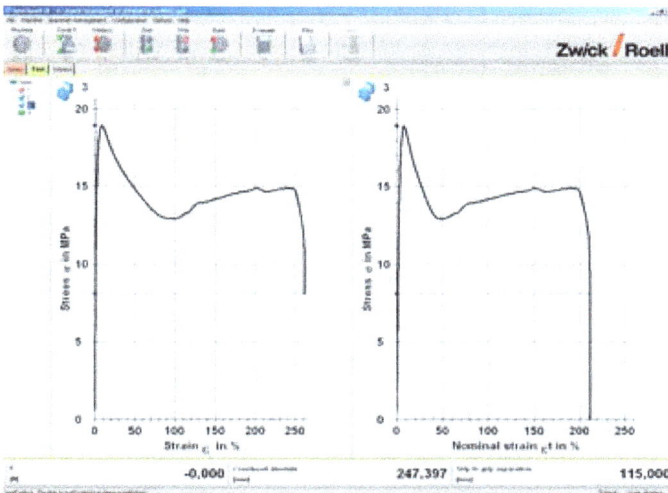

Figure 5.7 Graphs of tensile stress. Source: Zwick Roell

Different stiffness properties are obtained with the tensile stress test (see Figure 5.8):

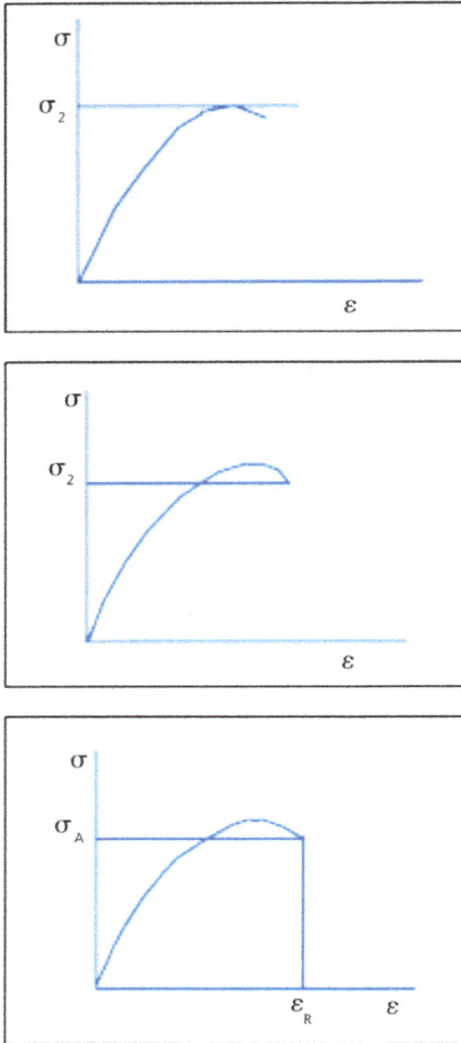

Figure 5.8
Stress–strain curves – top: tensile stress at yield; center: tensile stress at break; bottom: strain at break

Tensile stress at yield

Tensile stress at yield refers to the maximum stress level reached before material failure. It indicates the level of stress that can be applied to the material to cause it to deform without breaking. At this stress level the material will fail beyond repair.

Units: kg/cm^2, megapascal (MPa).

Tensile stress at break

Tensile stress at break is reached at the moment the material breaks. It is usually lower than the tensile strength at failure, due to the creep behavior of some plastics. At these stress levels the material breaks.

Units: kg/cm^2, megapascal (MPa).

Strain at break

Strain at break refers to the stretching (expressed in %) reached at the moment the material breaks. It indicates how much a material can stretch before breaking.

Units: %

5.1.5.2 Elastic and Tensile Moduli

During the tensile test, at the beginning of the stress–strain curve (Figure 5.9) we can see an area where the stretch is proportional to the tensile strength. The stress–strain of each material determines the slope of the curve. The tensile modulus is the relationship between stress and strain in the linear region, and therefore the region of direct proportionality.

During tensile deformation of a plastic material, there is a zone where the strain produced by the applied stress could be fully recovered when the stress ceases. Beyond a certain point of tensile deformation, it is no longer recoverable and considered permanent. The yield point is the stress level required to produce a permanent strain of 0.2%.

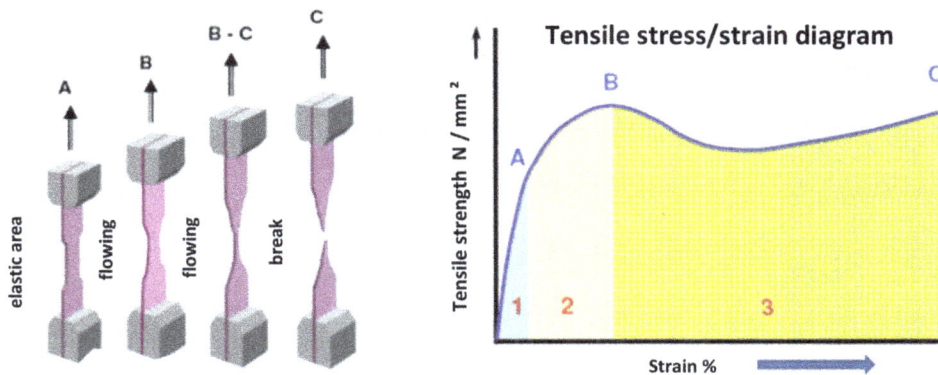

Figure 5.9 Elasticity of a tensile test specimen and a stress–strain graph

Explanation of zones in Figure 5.9:

- **Zone 1** (elastic zone): (reversible) deformation up to point A (proportional limit)
- **Zone 2** (stretching): the material is stretched as the pressure increases to point B (yield limit) = tensile stress at yield
- **Zone 3** (cold flow, creep of material): reduction or increase of the tensile strength to point C (tensile strength at break)

The stress data obtained from the tensile stress test can also be related to the ability of the material to withstand tensile stress during ejection or assembly after molding. During the ejection movement of the molded part, the material must withstand a high level of tensile stress; if the stress resistance of the material is exceeded, permanent deformation, breaks, cracks, etc. will occur, especially during movement of the slides, inner skids, and ejectors.

5.1.5.3 Impact Strength

Impact strength indicates the toughness of the material, its energy-absorption capacity, and its deformation. The values are given in kJ/m^2, kJ/cm^2, and J/m. There are two types of data (**notched** and **unnotched**) and two types of tests (**Izod** and **Charpy**); see Figure 5.10.

Impact Test Izod Notched

Vertical specimen position

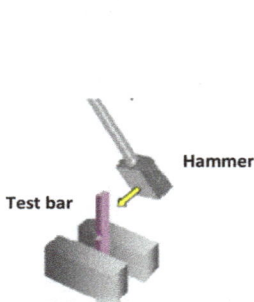

Hammer

Test bar

Impact Test Charpy Unnotched

Horizontal specimen position

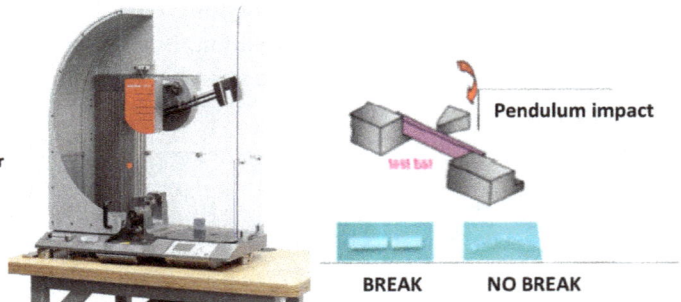

Pendulum impact

BREAK   NO BREAK

Figure 5.10 Impact strength: tests

5.1.6 Thermal Properties

Polymers exhibit a variety of thermal characteristics that define their behavior both in the injection molding process and their use in the final application when heat is applied. These characteristics can be characterized through different thermal tests.

5.1.6.1 Coefficient of Linear Thermal Expansion (CLTE)

The coefficient of linear thermal expansion (CLTE) is a measure of the expansion of a material (its length increases as the temperature increases by a certain number of degrees); see Figure 5.11. These values are important when different plastics are joined together, or when parts are assembled with metals or other materials that have different CLTEs.

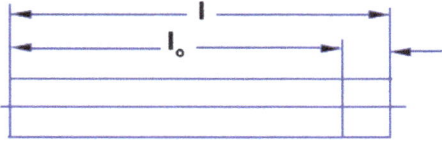

$$\alpha = \frac{\Delta l}{l_0} \cdot \frac{1}{(\vartheta - \vartheta_0)}$$

l_0 = length at temperature ϑ_0
l = length at temperature ϑ
$\Delta l = l - l_0$

Figure 5.11
CLTE calculation

5.1.6.2 Vicat Softening Temperature

The Vicat softening temperature is the temperature at which a plastic material softens rapidly. For semi-crystalline thermoplastics, the Vicat temperature is close to the melting temperature, while for amorphous thermoplastics it is close to the glass transition temperature.

The test is conducted by heating the test specimen and promoting penetration with a weighted needle (Figure 5.12). The Vicat temperature is the temperature reached when the needle penetrates 1 mm.

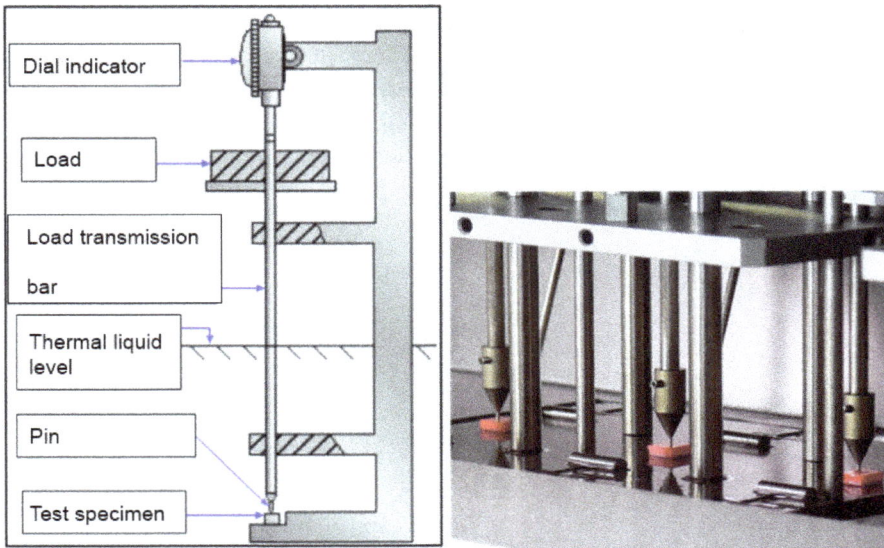

Figure 5.12 Vicat test diagram and Vicat test lab equipment. Source: ASCAMM, Zwick Roell

Vicat can be considered the limit for short-term use of a material. This value can also indicate the maximum temperature at which a part can be ejected by pressing the mold ejectors (Vicat < 10 °C) without permanent deformation of the part.

There are four different standard Vicat test conditions, the choice of which depends on the type of material; see Table 5.5.

Table 5.5 Standard Vicat Test Conditions

Vicat	Load (N)	Temperature (°C)
A 50	10	50
A 120	10	120
B 50	50	50
B 120	50	120

5.1.6.3 Heat Deflection Temperature (HDT/HDTUL)

The HDT is the temperature at which a polymer sample test specimen will deform under a specified load. This temperature indicates the ability of a material to function under stress for a short period of time.

By applying permanent stress and increasing the temperature of the specimen at a constant rate, we can see the effect of temperature on the stiffness of the material (see Figure 5.13). The HDT is the temperature at which the test bar or specimen (under load and temperature) reaches a deflection of 0.32 mm (ISO) or 0.25 mm (ASTM). This is the temperature at which the material has already failed, its properties have deteriorated, and it cannot maintain stiffness or tensile strength. The HDT temperature data also indicate the maximum temperature at which we can apply stress to a part during the injection process – for example, during part ejection.

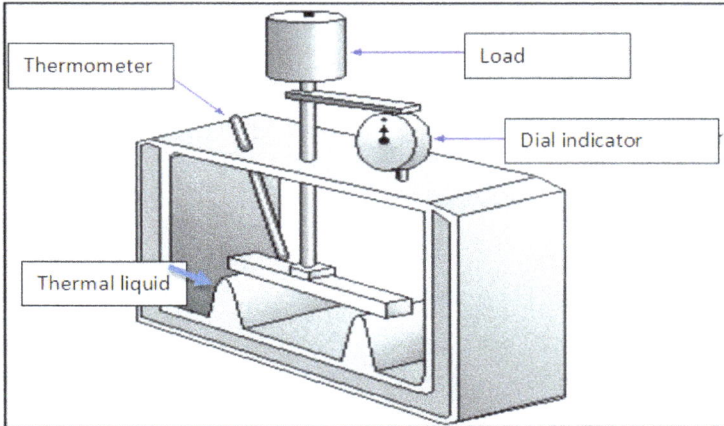

Figure 5.13 HDT test diagram and lab machine. Sources: ASCAMM, Zwick Roell

5.1.6.4 Critical Temperatures of Plastics

Table 5.6 shows critical temperatures for various plastics.

Table 5.6 Critical Temperatures of Plastics

Material	HDT (°C)		Vicat (°C)
	A, 1.8 MPa	B, 0.45 MPa	B, 50 N
PS	83	86	101
SB	83	91	95
SAN	86	95	99
ABS	91	98	101
PC	125–130	135–140	150
PC GF	145	150	155

Material	HDT (°C)		Vicat (°C)
	A, 1.8 MPa	B, 0.45 MPa	B, 50 N
PE	45	70	75
POM	85	150	155
PP	60	100	105
PA 6	95	190	200
PA 6 GF	200	200	215
PA 66	108	200	200
PA 66 GF	200	200	200

5.2 Part Information Checklist

In order to gather as much information about the part as possible, various checklists are used to consider aspects related to the end-use of the part. Although it may not seem important, it is essential to have as much information about the part as possible (its application, its requirements, environment, etc.) in order to be able to determine a robust injection process. Having as much knowledge as possible enables us to evaluate critical injection parameters, critical measurements, process tolerances, and so on. If we know the part's requirements, we can better determine the process conditions, process windows (aesthetic or dimensional), and the like.

A sample part requirements checklist can be seen in Figure 5.14.

Part requirements

Type of load
- Static
- Dynamic
- Cyclic
- Simple impact
- Repeated impact

Stress ratios
- Static
- Dynamic
- Compression, tension, bending stress amplitude

Deformation load
- Tension, compression, etc.

Apparent moduli
- Includes yield stress

Load direction

Safety factors

Size

Tolerance requirements
- Dimensional stability
- Coefficient of linear thermal expansion (CLTE)

Surface
- Surface hardness
- Coefficient of friction

Application environment – Product life

Ambient
- Humidity
- Water
- Chemical compounds
- Temperatures (minimum, maximum, and average)
- UV radiation
- Microwave radiation
- Sterilization
- Time

Safety factors

Aesthetic limitations

Shapes

Colors

Surface finishing: weld lines, parting lines, inlets

Decoration: printing, chroming, galvanizing, engraving

Costs

Current product cost

Amount of product needed for the chosen manufacturing process

Mold cost

Removing finishing operations

Redesign of part to simplify the product

Part weight

Price of material

Warehousing and distribution

Sector regulations

Flame resistance

Food contact approval

Nontoxic additives

Electricity

Military

Recycling

Figure 5.14

Sample part requirements checklist

5.2.1 Part Technical Specifications Checklist

Injection molded parts must meet stringent requirements relating to the final application for which they have been designed. Figure 5.15 shows a checklist of characteristics or properties usually required of plastic parts.

Specifications	Safety regulations	Oxygen limit
		Flammability (UL 94 classification)
		Food contact
		Medical criteria
		Contact with drinking water
		RoHS
	Sector regulations	Automotive (PSA, VW)
		Electronics (UL, CTI, glow wire)
		Appliances (RAL, UL)
		Medicine, packaging (FDA, BGA)
Environment	Resistance to chemicals	Solvents
		Water/humidity; chlorinated water
		Acids
		Gasoline, oil, etc.
		Detergents
		Stress-cracking
	Temperatures	High-low time
		Continuous heat resistance
		Expansion
		Vicat
		HDT
		RTI (with impact, without impact)
Structural properties	Resistance	Flexural modulus
		Tensile modulus (stiffness/flexibility)

Figure 5.15 Sample part technical specifications checklist

5.2.2 Target Factor Value Checklist

Molded parts often have to meet specific application sector regulations, standards as well as mechanical, thermal, and physical properties. Figure 5.16 gives a brief summary of the most frequent ones.

			Indicator	Regulations	Magnitudes
Specifications	Safety regulations	Oxygen limit		ISO	Earth's atmosphere (21.%)
		Flammability (UL 94 classification)	Combustibility index	UL94	HB-VOV1-V2-5V-5VB-5VA
		Food contact	European FDA, American FDA		
		Madical criteria	Pharmacopoeia, USP		
		Contact with drinking water	WRAS, NSF, KTW		
		RoHS			
	Sector regulations	Automotive (PSA, VW)	QK, TL		
		Electronics (UL, CTI, glow wire)			
		Appliances (RAL, UL)	RAL		
		Medicine, packaging (FDA, BGA)			
Environment resistance to chemical solvents					
		Water/humidity; chlorinated water			
		Acids			
		Gasoline, oil, etc.			
		Detergents			
		Stress-cracking			
	Temperatures	High-low time			
		Continuous heat resistance			
		Continuous heat resistance			
		Expansion	Coefficient of linear thermal expansion	ASTM D696	%
		Vicat	Vicat (10-50 N)	ISO 306	°C
		HDT	HDT (0.45-1.8 Mpa)	ISO 75	°C
		RTI (with impact, without impact)	RTI: Relative temperature index	UL 7468	°C
Structural properties	Resistance	Flexural modulus	Flexural modulus	ISO 178	MPa
		Tensile Modulus (stiffness/flexibility)	Tensile modulus, tensile resistance	ISO 527	MPa
		Impact	Impact strength (Charpy/Izod)	ISO 180-ISO 179	kJ/m^2
		Tensile Resistance	Tensile Resistance (break, yield)	ISO 527	MPa
		Flexural strength	Flexural strength	ISO 178	MPa

Figure 5.16 Sample target factor value checklist

5.3 Mold Information

The mold is the tool we use to produce parts by injection molding. It is essential to know as much information about the mold as possible (its design, properties, etc.).

It is useful to know the properties of the steel used to construct the mold:

5.3.1 Metals and Steels for Mold Construction

- Aluminum: used for prototypes and pre-production or initial production molds.

- Beryllium copper: less durable than steel and may scuff or wear faster than steel if used at the parting line. Beryllium copper can be used for inserts, slides or cores to increase heat transfer rates and reduce cycle time, due to its thermal conductivity.

- A2: steel hardened to 58-60 Rockwell C, wear resistant.

- D2: steel with higher chromium content, more wear resistant, more fragile, better for small parts, more difficult to machine than A2.

- D7: steel with high wear resistance but more difficult to process than A2 and D2.

- Stavax and Orvar: their chromium content and low porosity make them highly recommended for mirror-polished molds.

- 2311 steel, 2378 steel: baseplates.

- 2344 steel: removable and replaceable cores and cavities.

- 2510 steel: 52 HRC, cavities, cores, slides.

- P 20 steel: mold ejector plates.

- Nickel or chromium treatments (0.01 mm to 0.05 mm) give the mold surface better resistance to oxidation.

The choice of steel depends on the polymer and the quality and quantity of parts to be produced. However, the wide range of different steels makes steel selection difficult; the information in Figure 5.17 and Figure 5.18 is intended to help you choose the appropriate steel. Remember that for short production series and prototypes, materials such as aluminum or beryllium copper can be used.

W.Nr.	AUBERT & DUBAL	BÖHLER	SERMETAL	SIDENOR	STAHLMOL	THYSSEN IBERICA	UDDEHOLM
1.2080		K100	2080	ROK		THYRODUR 2080	SVERKER1
1.2083	X 1 3 T 6 W	M310	2083 ISO B	PLASTINOX	STM 1.2083	THYROPLAST 2083 ESR	STAVAX ESR
1.2085		M314	2316+5		STM 1.2317	THYROPLAST 2085	RAMAX S
1.2101		K245				THYRODUR 2101	
1.2210		K510				THYRODUR 2210	
1.2311			2311 ISO BM	EBRO 2311	STM 1.2311 STM SP300	THYROPLAST 2311	UDDAX
1.2312		M200	2312 ISO BM	EBRO 2312	STM 1.2312 STM SP300	THYROPLAST 2312	HOLDAX
1.2316		M300	2316 ISO B		STM 1.2316	THYROPLAST 2316	
1.2343	A D C 3	W300	2343 ISO B	FINOR	STM 1.2343 MAXIMUM	THYROTHERM 2343 EFS	VICAR SUPREME
1.2344	S M V 4	W302	2344 ISO B	FINOR V	STM 1.2344	THYROTHERM 2344 EFS	ORVAR
1.2363	S M H	K305	2363			THYRODUR 2363	RIGOR
1.2365	S M R	W320	2365 ISO B	MOLFOR		THYROTHERM 2365 EFS	QRO-90
1.2367	R 6110	W303	2367 ISO B			THYROTHERM 2367 EFS	DIEVAR
1.2379	SANCY 2	K110	2379 ISO B	ROK EXTRA	STM 1.2379	THYRODUR 2379	SVERKER 21
1.2380						THYRODUR 2380	VANADIS 4
1.2436		K107	2436			THYRODUR 2436	SVERKER 3
1.2510		K460	2510	MAGNO	STM 1.2510	THYRODUR 2510	ARNE
1.2550		K455	2550	CHOKER		THYRODUR 2550	REGIN-3
1.2581	VOLVIC	W100				THYROTHERM 2581	
1.2601		K105	2601			THYRODUR 2601	
1.2709			2709			THYRODUR 2709	
1.2711		W500	2711 ISO B		STM 1.2711 TDO	THYROPLAST 2711	
1.2714		W500	2714	ATOR 14	STM 1.2714 TDO	THYROTHERM 2714	ALVAR 14
1.2721		K605				THYRODUR 2721	GRANE
1.2738		M238	2378 ISO BM	EBRO 2738	STM 1.2738 STM SP300	THYROPLAST 2738	IMPAX SUPREME
1.2767	820 P	K600	SY67 ISO B	SUPERATOR		THYRODUR 2767	CALMAX
1.2842		K720	2842	EXTRAFORTE	STM 1.2842	THYRODUR 2842	
1.2886		W705				THYROTHERM 2885 EFS	
1.4980	R 25 T 2	T200				THERMON 4980	

Figure 5.17 Steel equivalences among different steel manufacturers. Source: Eurecat

Steel type	Abbreviated designation	Number of material	Tensile strength (kp/mm^2)	Hardness (HRC)	Notes
Hardened steel	14 Ni Cr 14	1. 5752	75	62–64	For medium and large molds. Easy machining. Good polishing.
Hardened steel	X 19 Cr Mo 4	1. 2764	80	62–64	Good machining and polishing.
Tempered steel	X 45 Ni Cr Mo 4	1. 2767	85	58–62	For small molds and complicated engravings.
Corrosion resistant steel	X 35 Cr Mo 17	1. 4122	80–95		Resistant to corrosion and to contact with acids
Nitrided steel	30 Cr Mo V 9	1. 8519	80–120		Molds with high dimensional accuracy.
Nitrided steel	34 Cr Al 6	1. 8504	80–100		Parts subject to wear and parts with high dimensional accuracy.

Figure 5.18 Steel types and information. Source: ETP Clot

5.3.2 Runners

Distribution channels or runners must be balanced so that the material follows the same path and design for all cavities (Figure 5.19). This results in uniform, balanced filling and pressurization of all cavities.

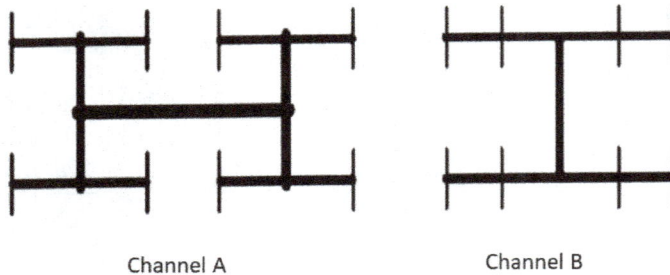

Channel A   Channel B

Figure 5.19 A) Balanced distribution channel; B) Unbalanced distribution channel

The distribution channel cross-section can have various shapes. The most suitable are circular and trapezoidal. In the latter case, it should be noted that the free passage cross-section will be in the shape of a circle inscribed in the trapezoid.

Cross-section of distribution channel: See Figure 5.20 for different runner cross-sections.

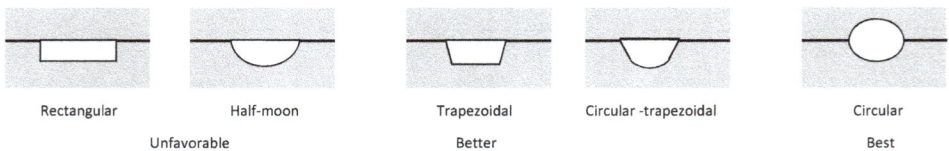

Rectangular	Half-moon	Trapezoidal	Circular -trapezoidal	Circular
Unfavorable		Better		Best

Figure 5.20 Different runner cross-sections

5.3.3 Cold Runner Design

For the appropriate pressurization of the runner system and a good flow path, the cross-section should be increased at every step or runner level. The thickness of the part (T) is the base value for calculating the entire runner system. This difference in channel size distribution allows the system to be properly pressurized and reach the cavity with the least-possible pressure loss.

For the proper runner channel sizing along the path of the molten material, the calculation begins with the part thickness – with each change in channel level, the thickness of the channel must be increased.

The key point is that the small sprue diameter should greater than the channel thickness that connects the gate to the part. To visualize this, see Figure 5.21, $D_2 > A$.

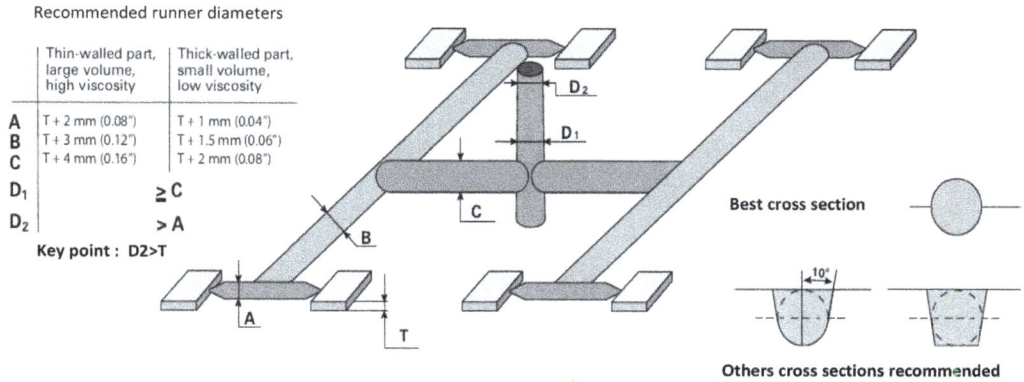

Figure 5.21 Cold runner thickness calculation for low- and high-viscosity materials. Source: DuPont

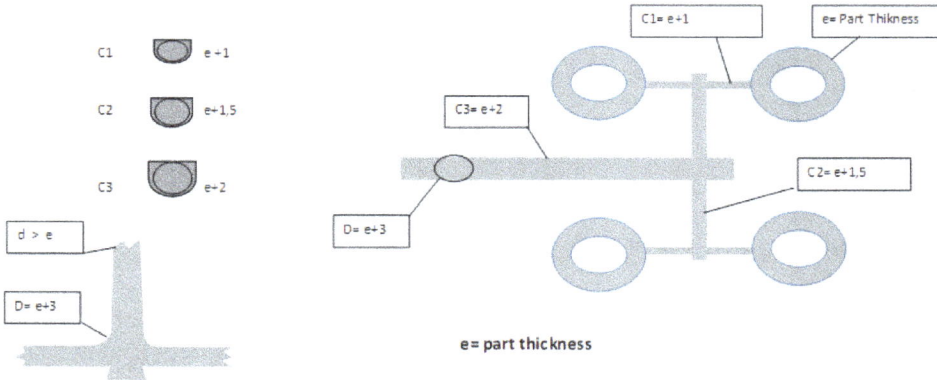

Figure 5.22 A cold runner calculation. Source: DuPont

5.3.4 Hot Runners

Changes of direction within the hot block in the hot runner system must ensure that no material is retained by the radii in the corners (Figure 5.23). Sharp edges or dead angles are potential promoters of material retention and degradation, which can cause aesthetic defects as well as a loss of mechanical properties.

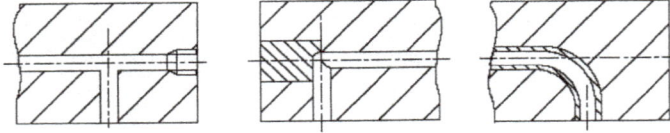

Figure 5.23 Left: inadequate; center: correct; right: perfect

5.3.5 Torpedo and Cavity Isolation

The torpedo is thermally insulated from the mold by the polymer because plastics have poor thermal conductivity. Possible degradation of this polymer can cause gases, bursts, and black spots due to the high temperature and long residence time. Alternatively, a hot runner tip, made of, e.g. DuPont Vespel® polyimide (Figure 5.24), can be positioned to isolate this area and avoid material retention, facilitate color changes, etc., which is also interesting for thermally sensitive or transparent materials.

Figure 5.24
Isolation of torpedo tip with Vespel® polyimide

5.3.6 Mold Cooling System Information

The cooling system controls the mold temperature. It is important to gather information about the cooling system in order to understand how it has been designed and be able to achieve the mold's maximum performance. Mold-temperature control is critical for obtaining parts without deformation and with high crystallinity.

The diameters of the cooling channels, the distance between channels, and the distance to the surface of the cavity are key points for an adequate cooling system (Figure 5.25).

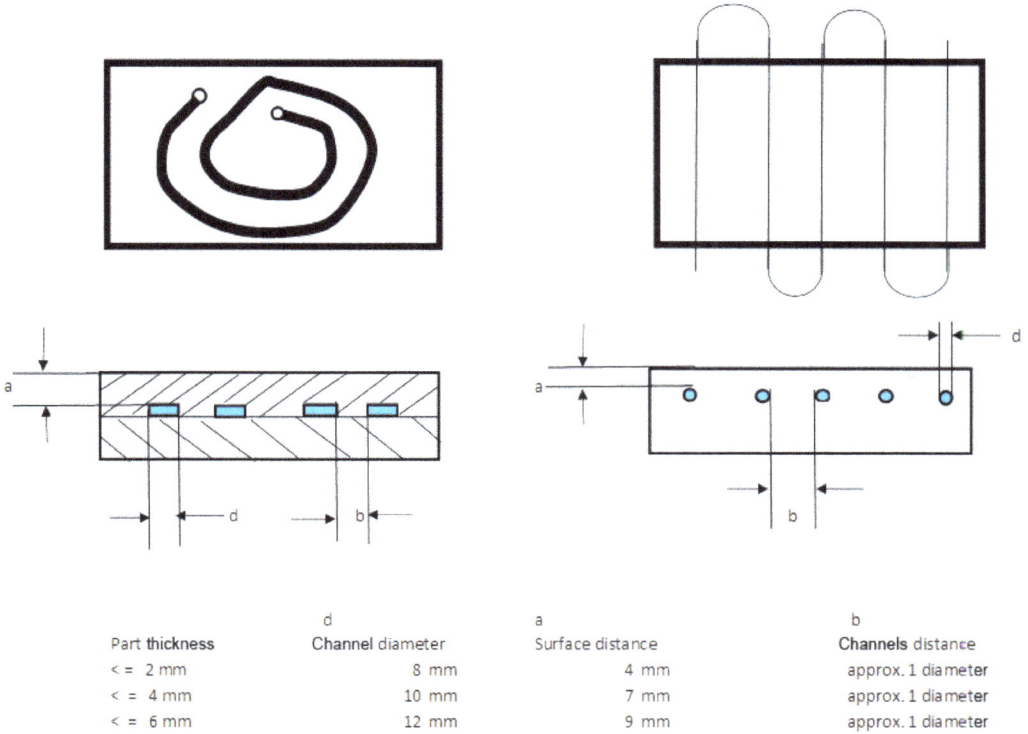

Figure 5.25 Design recommendations for mold cooling channels. Source: DuPont

Part thickness	d Channel diameter	a Surface distance	b Channels distance
< = 2 mm	8 mm	4 mm	approx. 1 diameter
< = 4 mm	10 mm	7 mm	approx. 1 diameter
< = 6 mm	12 mm	9 mm	approx. 1 diameter

Core Cooling Systems

Various types of cooling can be designed to provide coolant circulation through the cores; see Figure 5.26. Stereolithography systems are used to build 3D microchannel cooling systems that reach places unimaginable with conventional processing methods.

Core internal refrigeration system

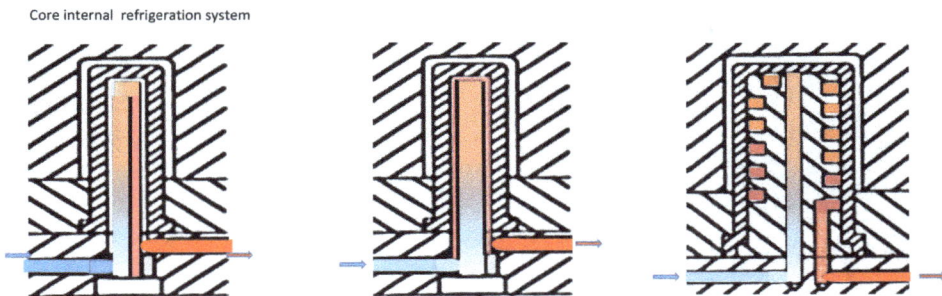

Figure 5.26 Different core-cooling system designs. Source: DuPont

5.3.7 Venting Information

Good and efficient mold venting is fundamental to a high-quality injection molding process. Venting channel dimensions such as depth, width, and length are critical to an efficient mold venting system (Figure 5.27). These dimensions depend on each material and its viscosity (Figure 5.28).

Venting in distribution channels and cavities   ▬▬▬ Venting channels

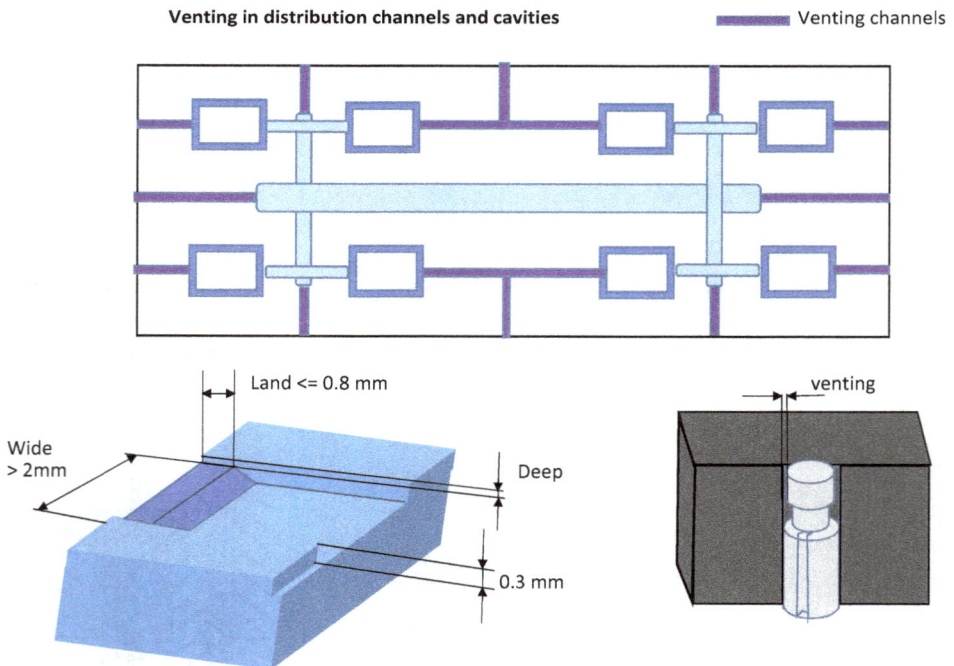

Figure 5.27 Venting design recommendations

Depth for semi-crystalline plastics	
Polyamide with glass fiber	< 0.03 mm
Ionomer	< 0.03 mm
LCP	< 0.01 mm
PBT	< 0.02 mm
Polyamide	< 0.03 mm
POM	< 0.03 mm
PP	< 0.025 mm
PPA	< 0.015 mm
PPS	< 0.01 mm
TPU	< 0.02 mm
TPV	< 0.025 mm

Depth for amorphous plastics		
Material	Easy flow	Reinforced with glass fiber
ABS	0.05	0.07
SAN	0.05	0.07
PS HI	0.05	0.07
PC	0.04	0.07
PSU	0.025	

Figure 5.28 Table of recommended dimensions for deep venting

Key points:

- The venting channel land should be between 0.8 mm and 3 mm. If it is too short, there will be a small steel surface area and it will not be kept open to airflow.
- If the length of the tunnel is excessive or its shape is inappropriate, volatile elements from the additive pack moving with the gases will leave residues in the venting channel, preventing the release of gases into the atmosphere.
- We need to vent 25–30% of the parting line or place a gas outlet for every 25 mm of parting line. Each outlet should be about 5 mm wide.

- The mirror-polished channels to atmosphere A1 are self-cleaning (SPI Finish A-1 -Grade #3, 6000 Grit Diamond).
- We must vent the distribution channels (Figure 5.29) in every 90° channel rotation. The venting channel for each runner should be as wide as the runner. The sprue center ejector should also be well vented.
- Deep venting in the runners: 0.03 mm for high-flow materials; 0.08 mm for medium-flow materials; and 0.12 mm for low-flow materials.

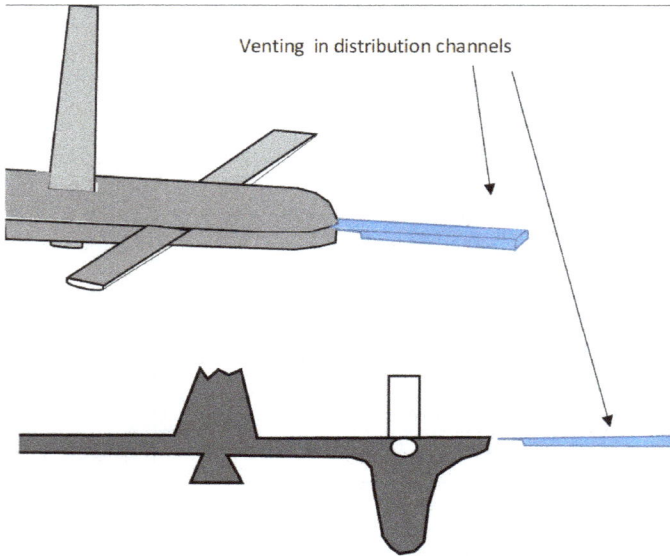

Venting in distribution channels

Figure 5.29
Examples of runner venting

Proper channel system venting prevents trapped air from circulating through the cavity. Each 90° turn of the cold runner, a generous gas outlet could be machined, since this is where we are not concerned about material that may cause flash.

Venting in Ejectors

Ejector pins are used in the mold venting system (Figure 5.30), especially in areas where venting through the cavity parting line is not possible.

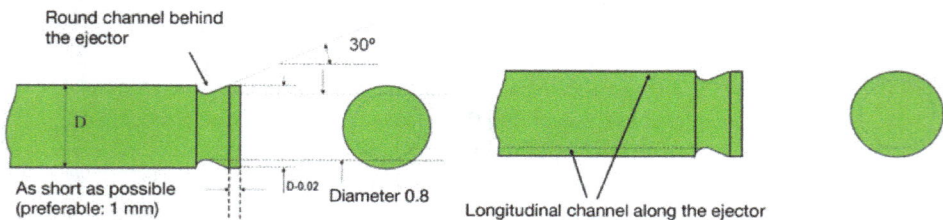

Round channel behind the ejector

30°

D

As short as possible (preferable: 1 mm)

D-0.02   Diameter 0.8

Longitudinal channel along the ejector

Figure 5.30 Recommended designs of venting ejectors. Source: DuPont

5.3.8 Draft Angles

The correct draft angles will allow us to remove the part from the mold without applying too much stress to the part and the mold. These draft angles depend on the shrinkage of each material and must allow the part to be easily removed from the mold (see Figure 5.31 and Table 5.7). There are occasions when the design of the part or its functionality does not allow for correct draft angles, in which case it is necessary to use slides and mechanical or hydraulic movements in the mold to demold these non-draft angle areas.

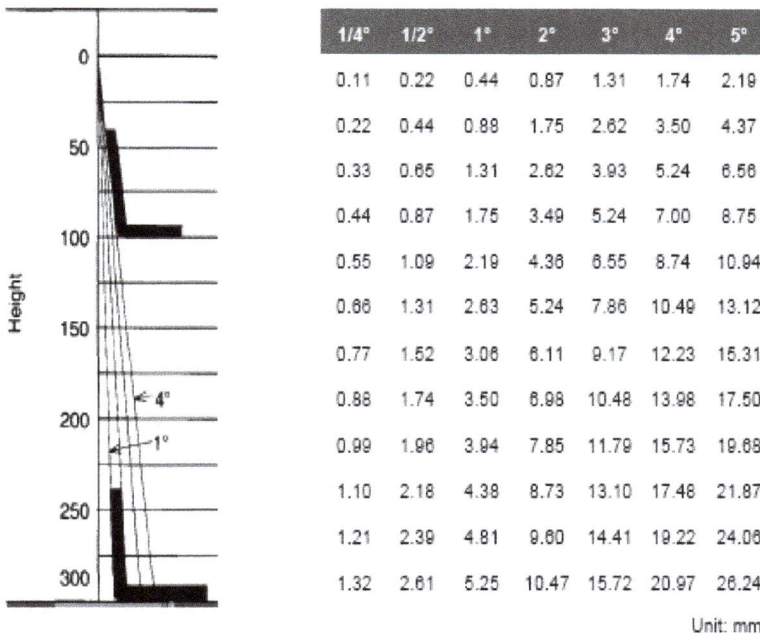

1/4°	1/2°	1°	2°	3°	4°	5°
0.11	0.22	0.44	0.87	1.31	1.74	2.19
0.22	0.44	0.88	1.75	2.62	3.50	4.37
0.33	0.65	1.31	2.62	3.93	5.24	6.56
0.44	0.87	1.75	3.49	5.24	7.00	8.75
0.55	1.09	2.19	4.36	6.55	8.74	10.94
0.66	1.31	2.63	5.24	7.86	10.49	13.12
0.77	1.52	3.06	6.11	9.17	12.23	15.31
0.88	1.74	3.50	6.98	10.48	13.98	17.50
0.99	1.96	3.94	7.85	11.79	15.73	19.68
1.10	2.18	4.38	8.73	13.10	17.48	21.87
1.21	2.39	4.81	9.60	14.41	19.22	24.06
1.32	2.61	5.25	10.47	15.72	20.97	26.24

Unit: mm

Figure 5.31 Demolding height as a function of angle and length. Source: DuPont

Table 5.7 Draft Angles for Different Plastics

Material	Recommended Draft Angle
ETPV Elastomer	1° to 2°
Ionomer	2° to 3°
LCP	0.2° to 0.5°
PA	0.25° to 1°
PBT-PET	0.5° to 1°

Table 5.7 Draft Angles for Different Plastics *(continued)*

Material	Recommended Draft Angle
PC	1° to 2°
POM	0.5° to 1°
PP	0.5° to 1°
PPA	0.5° to 1°
PPS	0.25° to 1°
TPC ET Elastomer	0.5° to 2°

Shrinkage Information

Polymer shrinkage in the part is complex to predict. Shrinkage is affected by many variables, such as the temperature of molten material, mold temperature, part thickness, cavity pressure (which depends on the gate size, runner, etc.), cycle time, possible reinforcements causing anisotropic contraction, etc. However, polymer manufacturers test samples and provide specific values for each grade and polymer family (Figure 5.32).

EVA	0.2% to 0.8%
GPPS	0.4% to 0.6%
LCP	−0.07% to 0.5%
PA 66	1.2% to 1.6%
PA 66 30% FV	0.3% flow 1.1% across
PBT	1.5% to 1.7%
PBT 30% FV	0.3% flow 1.1 across
PC	0.5% to 0.7%
PE	1% to 3%
PET	30% FV 0.2% flow 0.8% across
POM	1.9% to 2.2%
PMMA	0.4% to 0.7%
PP	1% to 2%
PP + FV	0.5% to 0.9%
ABS	0.6% to 0.9%
SAN	0.4% to 0.6%
ASA	0.3% to 0.8%
PP	1.3% to 1.6%
PA	0.25% to 1.5%

Figure 5.32 Shrinkage of various polymers

5.4 Summary of Required Information

By way of a summary and in addition to the information discussed in this chapter, we must not forget the following important information necessary for advanced injection molding:

- **Material**
  - Material manufacturer's injection molding recommendations
  - Recommended melting temperature range

- Melting temperature (semi-crystalline materials)
- Maximum recommended residence time at process temperature
- Recommended mold temperature
- Material drying temperature and time
- Maximum material tangential velocity
- Shear stress or maximum shear allowed for the material
- Material thickness to flow path ratio
- Recommendations for gates, channel design, etc.
- Maximum moisture (%) allowed in the material to be molded
- Technical data sheet, viscosity, MFI, Vicat, etc.
- Shrinkage, recommended draft angles
- Additives, content

- **Part**
  - Part volume
  - Part weight
  - Drawing tolerances
  - Original sample (if possible)
  - Use and operating environment of the part, application
  - Flatness, thickness, etc.
  - Number of cavities
  - Required part surface finishes

- **Mold**
  - Mold blueprint and design
  - Bottom and top plates, locate ring dimensions, etc.
  - Type of injection unit nozzle (conical, radial, etc.) and dimensions
  - Mold weight
  - Cavity gate position
  - Distribution runner channel system (layout, dimensions, etc.)
  - Movements, slides, hydraulics, etc.
  - Steel used (cavity, nozzles, etc.)
  - Cooling layout (detailed diagram, including dimensions)
  - Mold flow (if available)
  - Applied venting (location and dimensions)
  - Detailed diagram of electrical layout (connections, cores, etc.)
  - Kinematic mold movement
  - Applied surface treatments

- **Machine**
  - Machine specifications, plate dimensions, maximum and minimum mold sizes, etc.
  - Type of injection unit nozzle (conical, radial, etc.)
  - Maximum available injection unit temperature when setting
  - Screw diameter and type
  - Screw compression ratio
  - Injection unit L/D Ratio
  - Clamping force (Tn)
  - Maximum injection volume
  - Maximum injection pressure
  - Maximum injection speed
  - Injection unit intensification ratio
  - Switch over from filling to packing and holding (type, options)
  - Machine Delta P
- **Machine equipment**
  - Robot, range, transportable weight, etc.
  - Mold-temperature control, coolers, number, maximum temperatures, fluid type, flow rate in liters per minute, system pressure, pressure-flow graph
  - Dryer, volume, maximum set temperature, dew point temperature
  - Dehumidifier air filter type
  - Granulators (speed (rpm), power, granulometry)
  - Water coolers (power and flow rate)

6 Necessary Equipment for Advanced Injection Molding

6.1 Importance of Technical Equipment

Injection molding plants must be well equipped to define robust and productive processes, not only with injection machinery, robots, peripherals with reliability and quality, but also with elements that allow the process technicians to correctly apply scientific injection molding tools, for the definition of optimal process parameters and the analysis of possible process deviations. Some of these elements are helpful and complementary, while others are indispensable for a correct injection molding process definition.

6.2 Technical Equipment for Injection Molding and the Application of Scientific Injection Molding

Here is a list of some of the technical equipment that can greatly help injection technicians define a robust and productive injection molding process and, as previously mentioned, analyze process deviations and optimizations:

- Raw material, pellets, moisture meter or moisture analyzer
- Temperature probe
- Thermal infrared camera
- Balance
- Micrometer caliper gauge
- Shore A, Shore D durometer

- Microscope

- Surface tension measurement inks

- Polarized lenses

- Adjustable magnifying glasses

- Gauges, calibrated rods of different diameters

- Melt flow index tester

- Miscellaneous equipment

6.2.1 Raw Material, Pellets, Moisture Meter or Moisture Analyzer

This equipment is preceded by a long-standing argument. The most "scientific" method of measuring moisture in plastic pellets is carried out in the laboratory using the Karl Fischer system. This is a classic system used in analytical chemistry that uses a titration to determine the amount of water in an analyzed sample. It was invented in 1935 by the eponymous German chemist. This is a complex method that is difficult to apply in injection molding plants, which usually do not have an on-site chemical laboratory.

For this reason, there is an alternative that, in layman's terms, provides us with moisture data that can help us to assess the existing moisture in a sample of plastic pellets to check if the material can be processed well, namely thermogravimetry. In this, the sample is heated to a certain temperature and the weights of the sample before and after heating are compared, the assumption being that all the weight loss between the two samples is due to the evaporation of the moisture contained in the material sample being analyzed.

Here, critics of this system argue that some of the sample's additive package is also lost when pellets are heated during the test. This is true, but that can be minimized if the test temperature and duration are correct and not excessive. The test conditions must be sufficient to cause the contained moisture to evaporate, without causing any thermal degradation or loss of some of the additive package.

With this equipment, we can see the moisture levels in samples previously analyzed and so assess the efficiency of our drying system, system filter conditions and the efficiency of the drying time in relation to the temperature. In addition, the test is a fundamental tool in the analysis of process deviations should a molded part fail in its final application. Being able to determine the moisture level in the pellets at the time of molding the part allows us to find the cause of a loss of mechanical properties, such as impact or stiffness, or the root cause of potential aesthetic defects such as gloss, bursts, or streaks.

The test is also a tool that will help to efficiently manage and control a multi-dryer system. If we have several injection machines connected to the same dryer, we can determine and control if, at any moment during the production, the material consumption of the molding machine (the material output of the dryer) is higher than the maximum permitted. This means that the time that the material remains inside the dryer is shorter than is necessary for ensuring that the minimum moisture level in the pellets has been reached. The result is residual moisture problems in the pellets and therefore defects in the molded parts.

There are very practical pellet moisture analyzers for injection molding plants which are small and versatile enough to be recommended for good process control.

6.2.2 Temperature Probe

This is an essential piece of equipment in an injection molding plant. The temperature of the molten material is one of the important outputs that must be monitored, both for recording purposes when the process is approved and for checking in the event of any process deviations. It can also be used to measure the temperature of the mold, the cavity steel, and any other element that may require monitoring or a temperature measurement. In that case, the probe works by contact, not by penetration.

There are different types of melt temperature gauges, with the main one utilizing probe penetration (it is placed inside the molten material); while there are also non-contact laser temperature gauges, they are less accurate. The latter have the disadvantage of being affected by the angle of incidence of the laser beam in the measurement and, in the case of measurements on metallic and polished surfaces (e. g. the mold cavity), highly fluctuating readings can be obtained. In any event, we need to have a good, versatile temperature gauge for molten material and the mold cavity.

6.2.3 Thermal Infrared Camera

As a complement to the aforementioned temperature-measurement equipment, a thermal infrared camera is an excellent tool for managing and controlling the temperature of the injection molding process. By using a thermal camera, we can obtain a thermal image of the mold, the part, the cooling systems, and even the electrical power and electronic control panels. This represents an advance in the collection of process information and control over the main thermal outputs of the process. In fact, these cameras can even record a video that can be further analyzed (e. g. a mold opening and ejection sequence during an injection molding cycle can be recorded and the video can then be analyzed for part temperature during demolding, uniformity of part temperature and temperature differences between different parts in the same injection cycle).

Comparisons between cavities, part distortion, sink marks in specific areas, and the like can be analyzed in a much more scientific and accurate manner with this tool. One argument for investing in this equipment is that it can also be used for the preventive maintenance control of electrical circuits in machines and facilities, detecting "hot" areas before failure occurs.

6.2.4 Balance

This is an essential piece of equipment for defining an injection process and quality control. This equipment allows us to determine the exact holding pressure time on the basis of the gate freeze time, the mold filling balance level between cavities, and the repeatability of the process and the precision or repeatability of the injection machine, amongst others. For this reason, it is key to controlling and defining injection molding processes.

6.2.5 Micrometer Caliper Gauge

Whatever the sector and precision of the parts, we must have one of these tools to measure the critical dimensions of the parts (thickness, gate and runner dimensions, etc.). It is also essential for quality control of the production part series.

6.2.6 Shore A, Shore D Durometer

This equipment is used to measure hardness, normally elastomer hardness, on two scales: Shore A and Shore D. We can test the hardness of a great many elastomers and of some polyolefins. The durometer is suitable for measuring hardness in parts and comparing the hardness of different materials.

6.2.7 Microscope

Microscopes help technicians analyze part problems or quality deviations. By magnifying an image of the part under the microscope, we can identify the causes of some potentially surprising problems that are not so obvious to the naked eye.

We can choose from a wide variety of magnifications and qualities when it comes to microscopes. Capturing images that can then be exported to reports is recommended. Magnifications ranging from 40x to 80x are usually sufficient for the needs of an injection molding plant.

6.2.8 Surface Tension Measurement Inks

When a molded part is subjected to a surface-finishing process, such as painting, hot stamping, or screen printing, it is crucial to check its surface tension. The surface-tension measurement inks used for this also lend themselves to validating pre-treatments of molded parts, such as flame treatment and plasma treatment.

Inks of different surface tension are applied to the surface of the test part to determine its surface energy. They usually come in small pots with a brush. An alternative is to use marker pens. These contain an ink of a specific surface tension, which is indicated on the pen. A line is drawn on the part and the surface tension is then determined.

6.2.9 Polarized Lenses

It is well known that internal stresses in transparent molded parts can affect their properties and that these stresses influence the birefringence of the part in question. Therefore, we can see the different stress levels in the part through polarized lenses. The system I used consists in placing the part or surface to be monitored in the middle of two polarized lenses so that the internal stress levels in the part structure can be seen. It is interesting to be able to determine the molding conditions that lead to less stress in transparent materials, as well as to investigate incidents or process deviations.

This equipment can range from free-standing single lenses to illuminated holders in which the part is placed between the lenses.

6.2.10 Adjustable Magnifying Glasses

These are simple, practical analysis tools for revealing details that could go unnoticed, aesthetic defects, and surface blemishes on parts. They are more convenient to use around the machine than a microscope. A portable magnifying glass should always be in an injection molding technician's toolbox.

Typical magnifications for these portable and even pocket-sized lenses are 10x and up to 25x.

6.2.11 Gauges, Calibrated Rods of Different Diameters

These are useful for conducting measurements by inserting different calibrated rod diameters into circular gates or tunnel gates, as well as for quality control, to check critical hole diameters in molded parts.

6.2.12 Melt Flow Index Tester

Very interesting equipment for monitoring the melt flow index in cases such as:

- Reception of raw material:

  We can statistically monitor the material batches received in terms of the material's viscosity. It is well known that viscosity has a great influence on the injection process and the properties of the molded parts. For this reason, the quality control of this property is particularly relevant.

- Monitoring the level of process aggression in the molded part:

  The injection molding process is "aggressive" toward the material: we initially subject the plastic pellets to temperature and friction in the barrel, then we pass the molten material through a narrow hole (the one in the nozzle of the injection unit), before finally passing the material through narrow channels that end in a very narrow gate to the cavity where we subject it to changes in direction, part thickness, pressure, etc. The injection molding process causes an inevitable loss of molecular weight because there are molecules that break due to this high stress. This loss of molecular weight means a loss of material properties and, therefore, part properties.

  Molecular weight loss changes the original material viscosity to an extent depending on the amount of injection molding stress applied to the material. The greater the molecular weight loss (greater the process stress), the lower is the material viscosity of the molded plastic. Therefore, if we measure the flow index of the polymer from an molded part and compare the data to the flow index of virgin polymer that has not passed through the injection molding machine, we will have a differential flow index related to the molecular weight loss that has been caused by the injection molding process.

  Part failures and process deviations can be supported with this differential flow index analysis. It can also be used to compare two different process parameter conditions to determine which is the less "aggressive" condition from a material perspective and which will better maintain its properties in the molded part.

6.2.13 Miscellaneous Equipment

- Slide hammer

  These are essential tools for removing parts or pieces of a plastic part from the mold. Different sizes are available for different types of molds and parts.

- Handled mirror

  These mirrors are used to see inside deep cavities or hard-to-reach mold features. Be careful when looking inside a cavity with a hot runner or the injection unit through the barrel throat – ALWAYS use a mirror. Serious accidents have occurred when the molten material explodes, due to accumulated gas pressure inside the hot runner or the injection barrel. We must always use a mirror to look and check.

- Container for purges

  Containers can be used to isolate purges from room temperature and provide an accurate measurement of the melt temperature. Normally machined from a PTFE rod and with a handle.

- Zinc stearate

  A chemical additive that can help improve the sticking or adhesion of some material pellets and the molten polymer inside the screw and as it passes through the barrel throat. It can improve the movement of the screw during the metering stage, especially when the material granulometry or the design of the screws and hoppers are unfavorable or in the molding of recycling material with pellet shapes that cause caking.

- Slips or lubricants

  Chemical products with different properties, all designed to improve part demolding or the part friction against the cavity steel. They also improve some assembly operations. By reducing the friction coefficient between the plastic and the steel, we reduce the extraction stress. Be aware that these are products which can migrate to the surface of the part where they facilitate subsequent assembly on one hand but, on the other, can also cause the "halo" effect on the surface, due to the migration of the product to the surface.

▪ Prussian blue paste

Prussian blue paste is used to check adjustments in tools and molds. It has long been used to check the correct contact surfaces between two mechanical elements. In plastic injection molding, it helps to check mold adjustment areas, venting areas, flash areas, etc. It is essential for assessing certain situations where the mechanical adjustments of mold elements are being questioned.

▪ Clamping force indicator papers

Papers soaked with special inks which, upon compression (e. g. by applying the clamping force to the mold), penetrate the surface of the paper with different colors that vary with the pressure level applied. This provides us with a map of the compression distribution of the mold on a certain surface, which would cover the paper during the test carried out. This is an interesting way to identify inconsistencies in the parting lines or areas of the mold where such inconsistencies are suspected.

▪ Dry-ice cleaners

These are used for mold cleaning and maintenance equipment. The surface of the mold is well cleaned by dry ice pellets thrown under pressure onto the mold. They are particularly suitable for cleaning molds with surface deposits and molds with textured or polished surfaces that require special treatment for cleaning. This equipment minimizes the cleaning time, ensures the correct treatment of the surface of the mold without damage or scratches, and ensures a perfect state of cleanliness of the mold with each application.

Dry ice is solid carbon dioxide (CO_2) that sublimes at a temperature of $-78.5\,°C$. It is characterized by its sublimation property (i. e. change from a solid to a gas), meaning it does not pass through the liquid state and consequently does not moisten or wet the surface to be cleaned. The pellets used are normally between 1.7 mm and 4 mm in length.

▪ Portable flow meters

These meters can be connected in line with the cooling circuit whose flow we want to check. There are flow meters on the market that not only indicate the flow that passes through the cooling circuit, but also maintain a permanent flow once regulated. Other flow regulators maintain the adjusted flow and show us if the flow is in a turbulent regime. The Reynolds number can be also controlled.

▪ Portable thermometers

Portable thermometers can be installed in series in the cooling lines to control the input and output cooling-liquid temperature. They are versatile, as they can be installed in different circuits and molds for checking changes in the cooling input–output temperature.

- Portable dew point temperature meter

  These can be installed to check the dew point temperature data of material dryer equipment. Once the check is complete, they can be installed in other equipment to be checked. They are not necessary if our dryer equipment has a permanent dew point temperature measurement which can be checked on the dryer control. The lower the dew point temperature, the lower is the moisture content in the air used for drying the material.

- Radius gauge

  Radius gauges will help us to check the radii in the critical areas of the parts (i. e. distribution channels, inside the mold, etc.).

- Moisture barrier bags

  These bags are used to store material pellet samples for later analysis of the residual moisture content of the pellets in a moisture analyzer. The quality of the moisture barrier bag is important for ensuring that the sample remains isolated from ambient moisture and that the original moisture level of the pellets is not affected while waiting for the test to be carried out.

- Purge and injection unit cleaning materials

  These products help to clean the deposits that remain in the injection unit, screw, barrel, hot runner, and barrel nozzle. It is important to regularly use these cleaning products – do not wait to use them when the level of deposits is so high that only a deep clean, by disassembling and physically cleaning all the elements of the injection unit, would ensure proper cleaning of the system.

  Over the course of production days and injection cycles, residues build up in the injection unit which can remain stuck to the screw, barrel, check ring valve, nozzle, etc. Unless we undertake regular preventive cleaning of the injection unit, these residues will reach a level that cannot be removed with injection unit cleaning products at all. It will be necessary to disassemble and physically clean the injection unit.

- Other items

  Items that should be available in an injection molding plant:

  - Flashlight, stopwatch, cutter
  - Lighter, indelible black and white marker pen, tape measure
  - Ballpoint pen to quickly check radii

7 Tools for Scientific Injection Molding

The various tests, studies, or tools for use in scientific injection molding are explained below. Each has a different and specific objective, but the common goal is to define robust and consistent injection molding processes.

7.1 Preliminary Studies and Calculations

Before molding parts with a new mold for the first time, we can carry out a series of preliminary studies and calculations to obtain some initial parameters and limitations in advance. This calculation will also help us to determine the optimal size of the machine to be used.

7.1.1 Estimation of the Theoretical Clamping Force Required

As explained in Section 2.2.3, the clamping force F_c is calculated using the following equation:

$$F_c = \text{injection pressure} \times \text{projected area of the part}$$

Of the components of this equation, the projected area of the part is the easiest to calculate by applying geometric calculation; it the projected area is the area over which the plastic pressurized by the injection pressure will exert a force in the opposite direction to the clamping force (i.e. in the mold-opening direction). The most difficult to predict is the injection pressure, the equation's other component. This value for the required injection pressure into the cavity is more difficult to predict because it is greatly influenced by factors such as:

- Material viscosity

- Part thickness

- Flow path

- Part and mold design (gates, etc.)

Figure 7.1 (first introduced in Chapter 2) correlates some of these factors to obtain a value for the injection pressure to employ in the clamping force equation that is as close to reality as possible. As explained in Chapter 2, in order to use this figure as an aid in calculating the estimated clamping force required, it is necessary to know the flow path, the thickness of the part, and the type of viscosity of the material to be molded. The example in red shows the specific pressures for a flow path of 180 mm and a wall thickness of 1.5 mm, depending on the different material viscosities classified in scales A, B, and C.

The pressure obtained from the graph must be multiplied by the projected area of the parts to be molded to yield the total clamping force required:

$$\text{Clamping force} = \frac{(\text{injection pressure} \cdot \text{area})}{1000}$$

Where the required clamping force is in tons, the injection pressure is in kg/cm², and the area is the projected area in cm²

It is recommended to add 10% to the values obtained in the graph according to the material, thickness, flow path and material viscosity (A – high fluidity; B – medium fluidity; C – low fluidity).

In Figure 7.2, by entering the values in the green boxes, we can obtain the clamping force in different units.

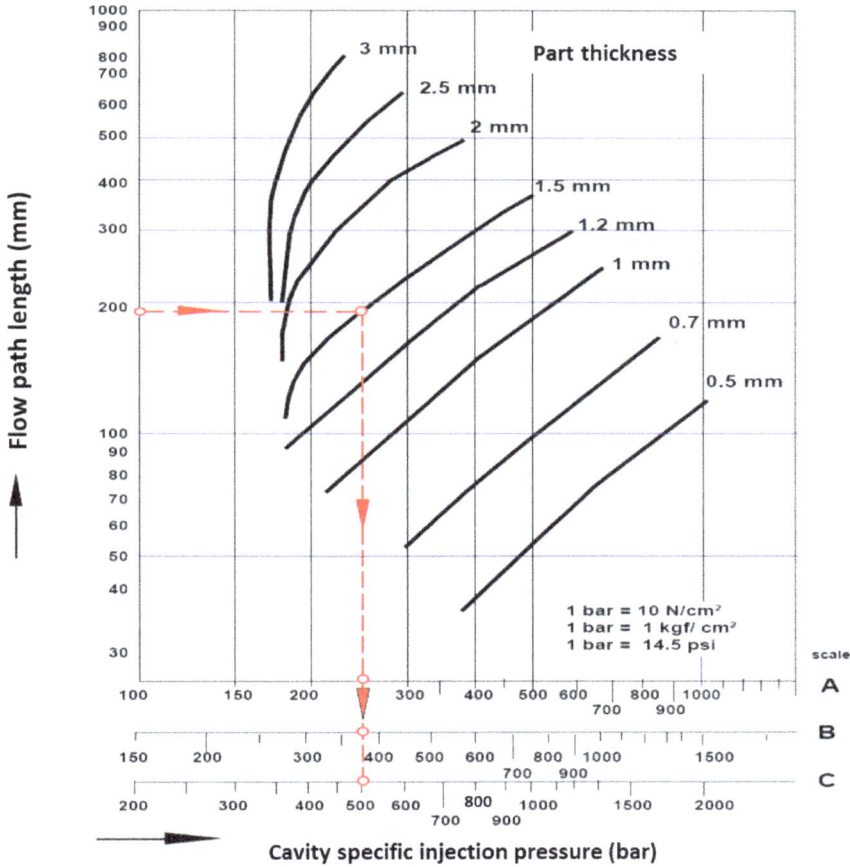

Figure 7.1 Graph for estimating cavity specific injection pressure. Source: Eurecat

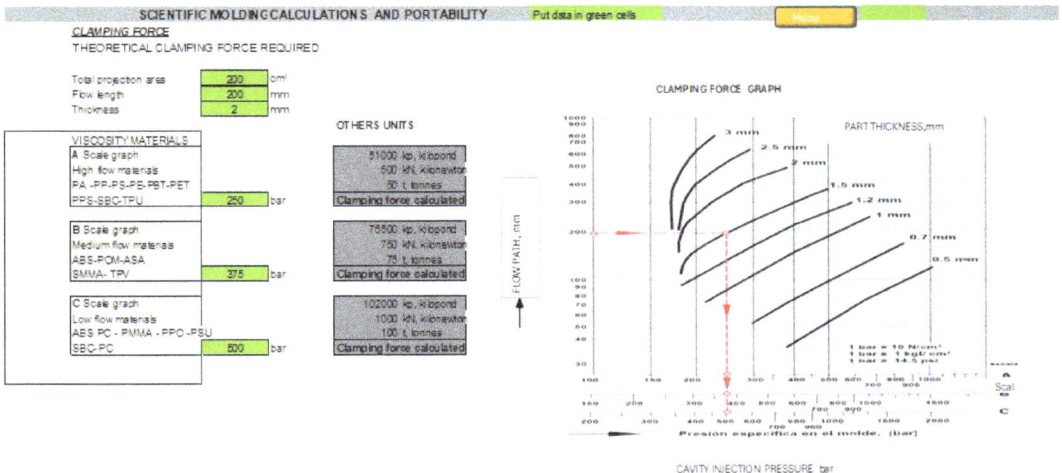

Figure 7.2 Example of a spreadsheet showing theoretical clamping force calculations

Machine and Mold Clamping, a Force and Area Ratio

Figure 7.3 Example of circumferential flash along the cavity parting line, due to insufficient clamping force

Important aspects to take into account in the mold clamping stage:

Sometimes when calculating the required clamping force, we may encounter the following situation. We know or estimate the peak of cavity filling pressure, we know the projected area or area on which the material will exert force in the mold-opening direction, we multiply these factors of pressure and area and get a theoretical clamping force. To this force we add a safety margin of 10% and we obtain a final theoretical clamping force value. It is true that in the molding industry we try to place a mold in the smallest-possible machine, for the sake of competitiveness, cost per hour, energy consumption, etc. We apply this calculated clamping force and, to our surprise, there are burrs or flash on the parts... What is happening?

When we analyze mold filling, we find that, inside the cavity, we are only applying some of the hydraulic injection peak pressure necessary to fill the mold. Here, there are some pressure losses due to, for example, the barrel nozzle, the hot runner or cold runner, and the cavity gates.

Upon further analysis of the mold filling, we must ask ourselves if we have considered all the projected area. If the mold design provides for slides, lifters or hydraulic cores (if these are locked in their closing position by mechanical systems that are kept locked by the machine clamping), the surfaces of these elements that will come in contact with the molten polymer should be taken into account because the cavity injection pressure applied to these surfaces will try to open the mold. More clamping force should be applied to these surfaces and so it must be considered necessary clamping force and added to the calculation of the theoretical clamping force required initially.

In the case of slides with draft angle locking, one way to calculate the clamping force to be added to this calculation would be to multiply the projected area or surface of the slide by the sine of the angle of the guide pin of the slide locking and multiply by the peak filling pressure. Another way to account for the required clamping force, due to slides, lifters, and the like, is to add to the projected surface area of the cavities 1/3rd of the projected surface area of the parts on the slides or lifters to the projected surface area of the cavities.

These forces that the slides will exert in the mold-opening direction should be added to the initial calculation to yield a more accurate clamping force calculation. There are molds that have a relatively small projected surface area but large slide surfaces that require large mold clamping force.

There are also authors who indicate that just closing the mold and keeping it aligned requires 10% of the total clamping force applied.

There are cases where the effort required to compress ejector plate movement springs, pneumatic dampers, and cases where different temperatures of each half mold to improve the demolding of mold zones or different coefficients of thermal expansion can cause an increase in the clamping force required for a given mold.

Let us analyze the clamping of an injection molding machine, which is the element responsible for keeping the mold closed during the whole process of filling and packing the material into the cavity. The machine, regardless of whether it is of the hydraulic, mechanical, or electrical clamping type (see Figure 7.4), if it has tiebars (on the market there are clamping injection molding machines without tiebars that offer excellent performance), these tiebars work as four large springs. When the mold is compressed by the clamping system, the tiebars stretch like a tensile spring and exert the clamping force that is exerted between the fixed platen and the moving platen and therefore on the mold that is contained between these two platens. This stretching of the tiebars is very important as it indicates the level of stress applied during closing and must be within acceptable values for the steel from which the tiebars are made.

In summary, when calculating the clamping force required, we must take into account, in addition to the projected surface area of the cavities, the projected surface areas of the slides that will also act during filling, and the force required to close and align the mold during locking. Allowance must also be made for springs or compression systems for mold movements, along with temperature differences between semi-molds and the force necessary to close and lock the mold.

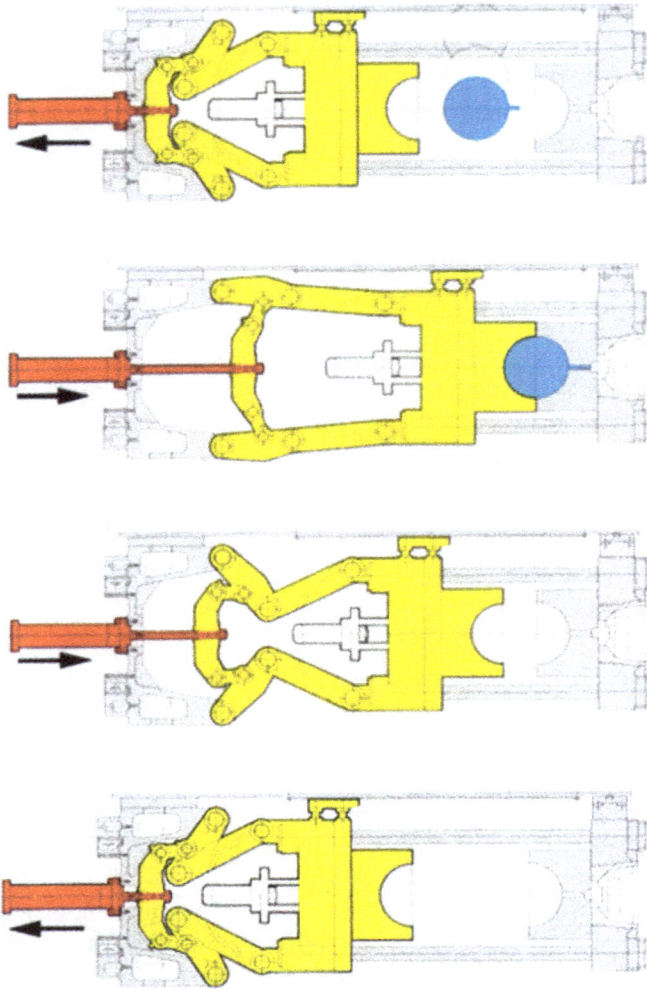

Figure 7.4 Sequence diagram of mechanical or toggle seals

7.1.2 Screw Rotation, Maximum RPM and Peripheral Screw Speed

Often, when molders talk about the screw speed used, some technicians refer to this parameter in terms of rpm, while others refer to it as a percentage of full rotation speed. Neither is valid from a scientific molding perspective. The correct measure for the screw speed parameter is the peripheral screw speed.

What is the peripheral screw speed? In kinematics, circular motion (also called cir-cumferential motion) is motion based on a constant axis of rotation and radius, where the trajectory describes a circle (see Figure 7.5). If we consider the tangential or pe-ripheral screw speed, the speed of a point on the outside of the screw diameter, then at the same rotation speed or rpm, this point on a screw diameter of 25 mm clearly will not travel at the same peripheral speed as on a screw diameter of 150 mm.

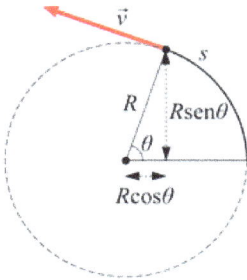

Figure 7.5
Peripheral speed diagram

In this case, one analogy would be a comparison between Big Ben in London and my wristwatch. If we look at both second hands for one minute, they will each have made one revolution after one minute has passed (i. e. 1 rpm), but both the distance traveled (given their diameter) and the speed at which the revolution will have been made will much greater in the case of Big Ben's second hand than in the case of my wristwatch. The second-hand on Big Ben will have completed the same number of revolutions as my watch, but at a much higher peripheral speed.

That is why, in scientific injection molding, we do not use rpm as a measure of screw rotation speed, but meters per second, for instance. There are still reputable plastic manufacturers who recommend maximum screw speeds in rpm!

Peripheral screw speed is an important factor in setting the parameters of the injec-tion molding process. We must not exceed the maximum peripheral speed allowed and recommended by the material manufacturer (see Table 7.1). This is often a factor that is not considered when setting process parameters. I have often observed – in injection molding plants – injection machines metering at very high screw unjustifi-able speeds because, after when metering was finished, there was a long wait for the cooling time to end.

During the metering phase, much of the energy required to melt and homogenize the melt plastic is generated by the rotation of the screw. Due to its design, this rotation or peripheral speed provides energy that supports the plasticizing process; in some cases, up to 80% of the thermal energy required to melt the polymer is generated by the rotation of the screw. However, this energy can be excessive and cause molecular breakage, with consequent loss of mechanical and aesthetic properties.

Tangential speeds higher than those recommended can lead to degradation and loss of material properties. In addition, they directly affect the following:

- Melt temperature, uncontrolled increases
- Plastic reinforcements and additives may be degraded
- Injection unit, barrel, and screw wear

Table 7.1 Recommended Generic Peripheral Screw Speeds (m/s)

Material	Maximum Peripheral Screw Speed
PE	0.8
PP	0.7
PS	0.7
PA	0.5
POM	0.1 to 0.25
PET	0.3
PBT	0.35
ABS, ASA	0.5
SAN	0.55
PC	0.5
PMMA	0.35
CA	0.45
PPE/PA NORYL	0.4
HYRTEL	0.4
ABS/PC	0.2
PA 66	0.8
TPU	0.2

The peripheral speed (m/s) is converted to rpm as follows:

$$\text{Max RPM} = (\text{Max velocity(m/s)} \times 60,000)/(2 \times \pi \times r)$$

See also Figure 7.6.

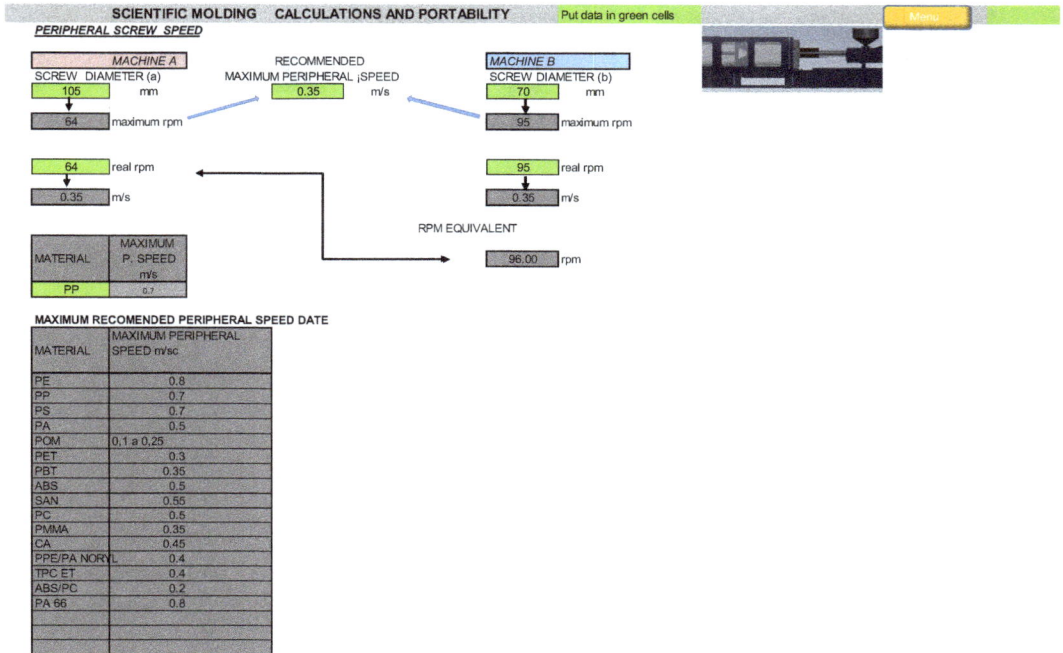

SCIENTIFIC MOLDING   CALCULATIONS AND PORTABILITY   Put data in green cells   Menu

PERIPHERAL SCREW SPEED

MACHINE A	RECOMMENDED	MACHINE B
SCREW DIAMETER (a)	MAXIMUM PERIPHERAL ¡SPEED	SCREW DIAMETER (b)
105 mm	0.35 m/s	70 mm
64 maximum rpm		95 maximum rpm

| 64 real rpm | | 95 real rpm |
| 0.35 m/s | | 0.35 m/s |

RPM EQUIVALENT

96.00 rpm

| MATERIAL | MAXIMUM P. SPEED m/s |
| PP | 0.7 |

MAXIMUM RECOMENDED PERIPHERAL SPEED DATE

MATERIAL	MAXIMUM PERIPHERAL SPEED m/sc
PE	0.8
PP	0.7
PS	0.7
PA	0.5
POM	0,1 a 0,25
PET	0.3
PBT	0.35
ABS	0.5
SAN	0.55
PC	0.5
PMMA	0.35
CA	0.45
PPE/PA NORYL	0.4
TPC ET	0.4
ABS/PC	0.2
PA 66	0.8

Figure 7.6 Example of a spreadsheet for converting rpm to m/s and calculating the rpm setting from the maximum peripheral screw speed recommended for each material (by entering the values into the green cells, we obtain the calculated values)

7.1.3 Metering Stroke or Volume Calculation

In order to estimate the theoretical machine setting for the metering stroke or volume at the beginning of an injection molding process definition, we need to know the following values:

- Screw diameter
- Part weight or volume
- Number of cavities
- Density of the molten polymer

We must bear in mind that, when we set a metering stroke in millimeters, we are actually defining a theoretical cylinder with a diameter equal to the machine screw or barrel diameter and a length equal to the metering stroke setting. With these values for the diameter and length, we can calculate the volume of the defined metering cylinder (see Figure 7.7). The result of the calculation must be increased by the desired cushion stroke value.

SCIENTIFIC MOLDING CALCULATIONS AND PORTABILITY   Put data in green cells

THEORETICAL DOSAGE STROKE OR VOLUME   DOSAGE

MACHINE A   MACHINE B

SCREW DIAMETER   MELT DENSITY   SCREW DIAMETER
85   mm   0.92   g/cm³   60   mm

300   g   299.97   300.00   g   300   g
57.46 mm dosage 57.46   115.33 mm dosage 115.33 mm dosage
326.09 cm³ dosage 326.06   326.09 cm³ dosage 326.09 cm³ dosage

	g/cm³ Density Melt materia
MATERIAL	
PA 6	0.96 POLYAMIDE 6

MELT DENSITY TABLE

MATERIAL	MATERIAL NAME	g/cm³ Density Melt material	g/cm³ Density 23 °C
ABS	ACRYLATE BUTADIENE STYRENE	0.92	1.05
ABS.PC	ABS -POLYCARBONATE	0.95	1.13
PA 6	POLYAMIDE 6	0.96	1.13
PA 66	POLYAMIDE 6-6	0.96	1.13
PA 66 GF 35	POLYAMIDE 6-6 -30% CLASS FIBER REINFORCED	1.15	1.41
PBT	POLYBUTYLENE TEREPHTHALATE	1.07	1.3
PBT GF 30	POLYBUTYLENE TEREPHTHALATE -30% GLAS FIBER	1.36	1.65
PC	POLYCARBONATE	1.04	1.2
PC GF 30	POLYCARBONATE 30% GLASS FIBER REINFORCED	1.3	1.44
PC /PBT	BLEND POLYCARBONATE ,POLYBUTYLENE TEREPHTHALATE	1.04	1.22
PE HD	HIGH DENSITY POLYETHYLENE	0.72	0.957
PE LD	LOW DENSITY POLYETHYLENE	0.7	0.917
PMMA	POLYMETHYL METHACRYLATE	1.04	1.19
POM	POLYOXYMETHYLENE	1.15	1.42
PP	POLYPROPILENE	0.73	0.905
PP T 20	POLYPROPILENE 20% TALC REINFORCEMENT	0.87	1.04
PP T 40	POLYPROPILENE 40% TALC REINFORCEMENT	0.98	
PP 40%FV	POLYPROPILENE 40% GLASS FIBER REINFORCEMENT	0.85	
PS	POLYSTYRENE	0.92	1.05
PVC	POLYVINYL CHLORIDE	1.15	1.43
SAN	STYRENE ACRYLONITRILE	0.99	1.08
PPO	POLYPHENYLENE OXIDE	1.06	1.1

Figure 7.7 Example of a spreadsheet showing theoretical metering stroke or volume calculations (by entering the values in green cells, we obtain the calculated values)

7.1.4 Theoretical Cooling Time

The calculation of the cooling time is one of the most complex calculations in the injection molding process. The calculations done by engineering molding simulation systems take into account factors such as the thermal conductivity of the molded polymer, or its ability to cool, and the specific heat of the polymer or its ability to melt. They also consider the thermal conductivity of the mold steel, the diameter of the cooling circuit, the distance between the coolant passages and the distance between the coolant passages and the cavity surface, the coolant flow circulating, etc. In short, cooling time is affected by many complex variables.

One of the most important factors in cooling time is the part thickness; small changes in thickness significantly alter the cooling time, due to the poor thermal conductivity of plastics in general. A simplification of the complex cooling time calculation is:

$$\text{Cooling time} = \text{thickness}^2 \times K$$

where K is a material-specific constant

The result of this calculation is a required cooling time that we can consider as an initial value in the testing and validation phase of the molding process (see Figure 7.8). It should be noted that this is an approximate calculation.

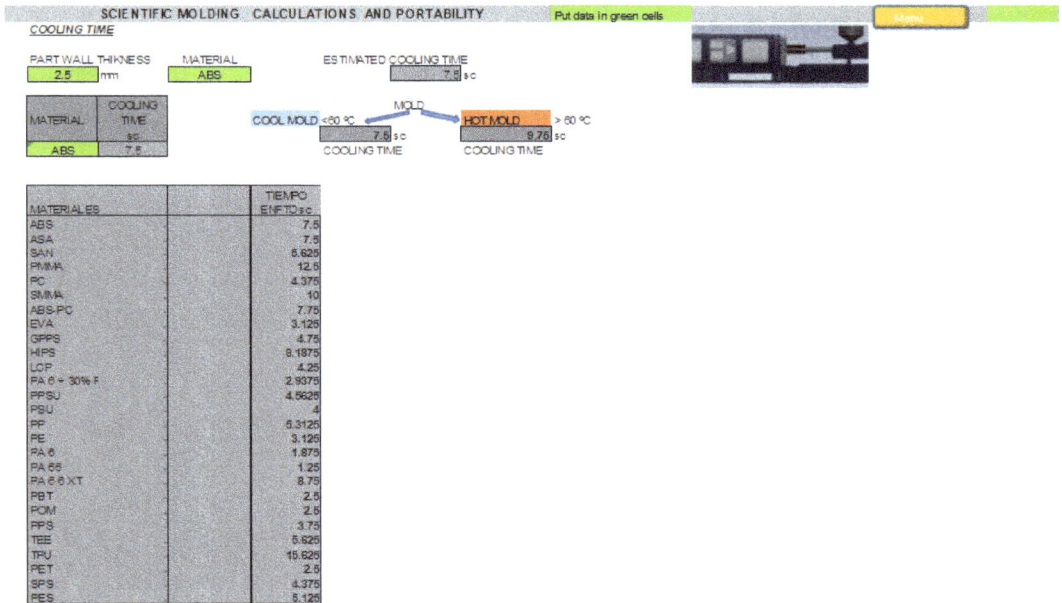

Figure 7.8 Example of a spreadsheet showing theoretical cooling time calculations (by entering the values in green cells, we obtain the calculated values)

7.2 Scientific Injection Molding Tools for Defining the Injection Molding Process

To properly define a plastic injection molding process by applying the methodology of scientific injection molding, the tools or studies explained in this chapter can be used. Each has a specific objective and addresses key process parameters, but they all share a common goal: the definition of robust and consistent (and therefore productive) injection molding processes.

7.2.1 Viscosity Curve Test or In-Mold Rheology Test

Introduced in 1980, this was initially not very well received by injection molders, although they are now using it when applying the scientific injection molding methodology. It was originally developed to demonstrate the non-Newtonian behavior of plastics and to prove that the viscosity of the melt is strongly affected by the applied shear rate.

7.2.1.1 Plastic Behavior

As explained in Section 4.3.2, a Newtonian fluid is a fluid whose viscosity is not affected by the applied shear rate. The viscosity is constant and does not depend on the shear rate. A non-Newtonian fluid is a fluid whose viscosity is affected by the applied shear rate. In the case of thermoplastics, these are non-Newtonian fluids and their viscosity is affected by the applied shear rate.

As shown in Figure 7.9 (first introduced in Chapter 4), as the shear rate increases, the plastic viscosity decreases, and conversely, as the shear rate decreases, the plastic viscosity increases. In the injection molding process, the shear rate is mainly applied by the cavity filling speed or the injection speed.

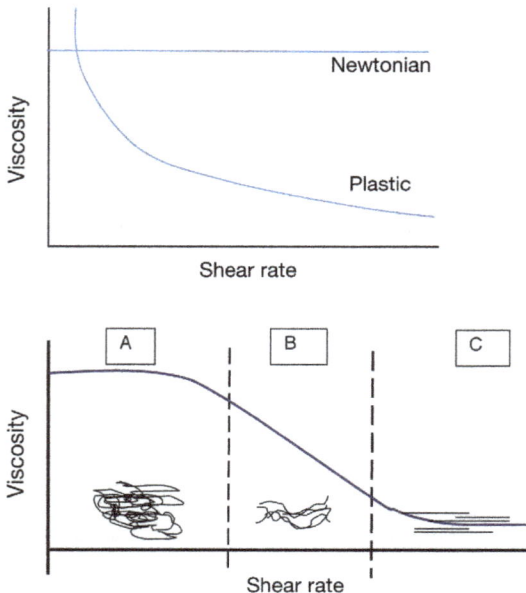

Figure 7.9
Differences in Newtonian and plastic behavior and viscosity development at different shear rates

What we are about here is generating and collecting a series of data to build a curve that shows us the viscous behavior of a plastic as a function of shear rate or injection speed. By conducting this study, we can determine the viscous behavior of the injected material in the molding process, in our injection molding machine, and with our mold. With the viscosity curve, we can decide which injection speed is the most suitable to focus the process on and make it more robust. The viscosity curve can also be used to compare the behavior of two "countertypes" or similar materials under real injection conditions, or raw materials from different supplier batches, and not just the values that appear in the material data sheet or in the COA (certificate of analysis) or quality certificate.

Viscosity graphs can be provided by the raw material manufacturers which obtain them through tests in their laboratories. Normally plotted on a logarithmic scale (see Figure 4.8), they show the decrease in viscosity values (Pa s^{-1}) as a function of the shear rate (s^{-1}) applied. Other information provided by these curves (e.g. in Figure 4.8) is that we can see, that at low shear rate values (500 s^{-1} to 1000 s^{-1}), changes in plastic temperature (210 °C, 215 °C, and 230 °C) have a much greater influence on viscosity than at higher shear rate values (6000 s^{-1} to 7000 s^{-1}). However, the advantage of performing the test on the injection molding machine is that the behavior of the "trinomial" machine, mold, and material can be observed in our facilities where the parts will be molded.

The data for constructing the relative viscosity curve are provided by a test called the relative viscosity test or in-mold rheology test.

7.2.1.2 Viscosity Curve (or In-Mold Rheology) Test

This test method and study allows us to observe the effect of injection speed on material viscosity, given the non-Newtonian behavior of plastic under an applied shear rate. To determine the optimum injection speed range, we can perform a relative viscosity test to obtain a viscosity curve.

With this test, we can see the behavior of the material–mold–machine set from the perspective of the plastic's viscous behavior. We can choose the injection rate necessary to maintain the viscosity as constant as possible or in the most Newtonian area of the material viscosity curve, so that the process can absorb and self-adjust to variability in plastic viscosity (different batches) or different injection speeds. The method is based on the possibility of separating or disconnecting the filling and packing and holding phases of the material in the cavity and carrying them out independently.

The viscosity curve is plotted from the data obtained when the mold is approximately 90% filled at different injection speeds and depends on the characteristics of the mold material and the injection machine. In this test, the real injection pressure value and the filling time at switchover point are therefore collected under the conditions of each shot for the different injection speeds.

With these values, it is possible to calculate the points on the graph that show the viscosity as a function of shear rate (Figure 7.10). Normally, the graph for plastics is L-shaped. The zone of optimum injection speed is in the most stable and flattest portion of the curve (near the "elbow") because the plastic exhibits Newtonian behavior in this zone.

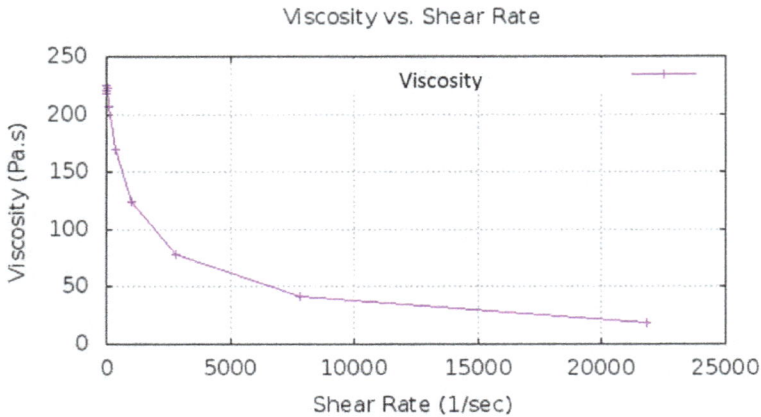

Figure 7.10 Viscosity as a function of different injection speeds or shear rates

By molding in the most stable and flattest portion of the curve, we ensure that the process will be less affected by variations in plastic viscosity, injection speed, or injection pressure (i. e. certain variations in injection speed and/or injection pressure will not seriously affect the material viscosity and, vice versa, certain material viscosity variations will not affect mold filling). This gives rise to a highly consistent and repeatable process.

During injection, the thermoplastic is subjected to various stresses of different magnitudes as the cavities are filled. If the viscosity value is on the curve's vertical slope, due to the stress applied by the filling speed, small variations in injection speed will be greatly affected by the viscosity of the material and vice versa. However, if speed and shear rate are used to obtain the viscosity values on the flat portion of the curve or Newtonian area, we will have a more robust and stable process.

7.2.1.3 Relative Viscosity or In-Mold Rheology Test: Injection Machine Configuration

1. Set the melt temperature in the machine and check that the melt temperature is the one recommended by the raw material manufacturer.

2. Set the machine with a holding pressure and holding pressure time of zero.

3. Set a high injection pressure limit so that it neither affects nor limits the process.

4. Set a cooling time to ensure that the parts are solid when demolded.

5. Set the switching (from filling to holding) system by stroke or volume.

6. Ensure that the switchover point is somewhat premature as it could cause flash and damage the mold when tests are conducted at high speeds (it could be 90% of the total volume, for example).

7. Set the lower test injection speed.

8. Proceed with a regular increase in the injection speed setting – the objective is to test and carry out shots at at least 10 different speeds between the minimum and maximum available machine injection speed.

9. In each cycle at different injection speeds, we must collect and record the data corresponding to the maximum injection pressure at the switchover point and mold filling time. With these data, we will prepare the table for the calculation of the relative viscosity or in-mold rheology (e. g. Table 7.2).

10. It is recommended that several injected shots be made at each speed change in the tested range to allow the machine to stabilize the data.

11. Using the values obtained and reported in the table, we can construct the in-mold rheology curve (Figure 7.11).

Table 7.2 Example of Test Data Collected for the Viscosity Test or In-Mold Rheology

Speed (mm/s)	1/Filling Time (1/s)	Filling Time (s)	P, Injection Pressure (bar)	s, Shear Rate (1/s)	Viscosity (Injection Pressure bar × Filling Time s)
15	0.21	4.80	64.00	0.21	355
20	0.36	2.80	70.50	0.36	197
25	0.63	1.60	74.30	0.63	119
30	0.71	1.40	76.30	0.71	107
35	0.83	1.21	78.80	0.83	95
40	0.94	1.06	82.80	0.94	88
45	1.03	0.97	84.90	1.03	82
50	1.11	0.90	86.00	1.11	77
55	1.18	0.85	88.90	1.18	76
60	1.27	0.79	91.00	1.27	72
65	1.32	0.76	91.70	1.32	70
70	1.43	0.70	91.90	1.43	64

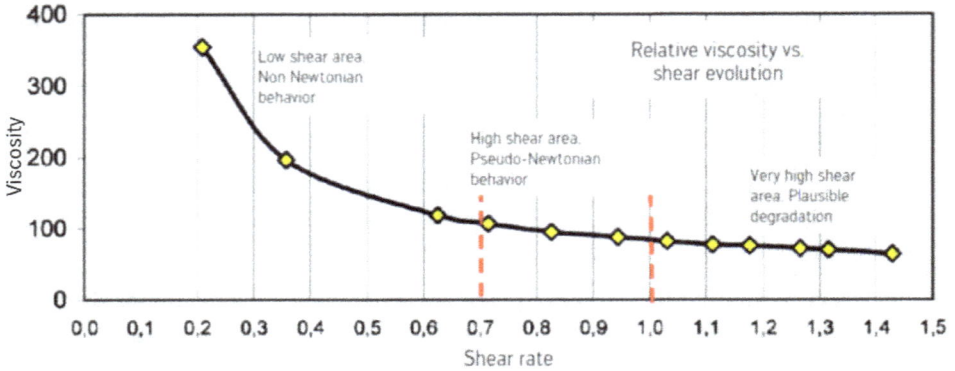

Figure 7.11 A viscosity data test table and its resulting viscosity curve

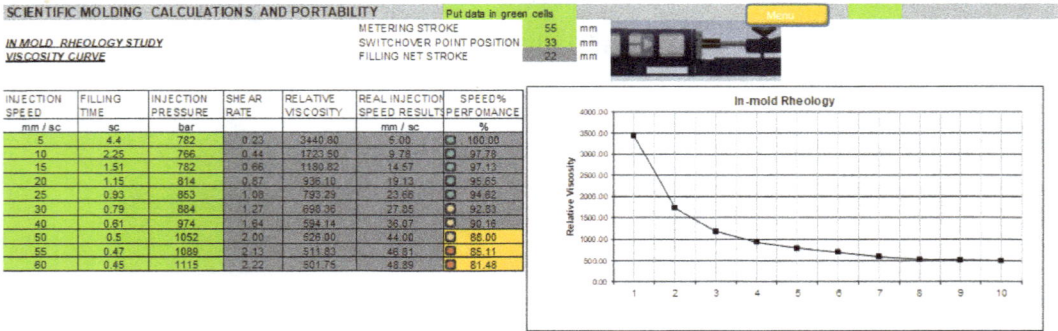

Figure 7.12 Example of a spreadsheet for conducting the viscosity test (by entering the values in green cells, we can obtain the calculated values and the viscosity behavior graph)

The elbow-shaped curve shows that at low shear rates, the viscosity is highly dependent on the injection speed (the vertical area of the graph). In this vertical area, small changes in speed will significantly affect the viscosity of the material and vice versa. However, at high speeds (i. e. the flat portion of the curve), the effect of the injection speed on the viscosity is minimal. We can conclude that the injection process is more robust and constant when injecting at repeatable and constant high speeds.

So why don't we always inject at the maximum injection speed available on the injection machine? Or why do we have five or ten injection speeds available to be set when only one at the maximum would be enough? Some of the many limitations that our process may have are: problems with the mold, venting, burrs, adjustments, type of part, sector, etc., not to mention the fact that a high injection speed can damage the plastic material and/or its additives through excessive shearing. As for high injection speeds, we could also have an important limitation regarding the injection molding machine – the machine that can reach the high injection speed setting.

So we have, as noted, material, machine and mold limitations that force us to set an injection speed that, as recommended, **should not be a speed with a relative viscosity in the vertical portion of the curve and should be a speed that implies a constant viscosity in the flat portion of the curve**, but also considers the aforementioned material, machine and mold limitations. This chosen speed will allow us to inject with a robust process that will have a constant and repeatable viscosity because in this portion of the curve, small changes in speed will not affect viscosity and conversely, changes in viscosity will not affect speed.

In summary, if we take the time to do this test in the first mold trials, we will learn more about the process (the mold, machine, material, etc.) in a one-hour test than in months of molding parts. There may be occasions where the injection speed obtained through the relative viscosity method is not applicable to the process (e. g. filling with a long flow path or with large thicknesses and optical and aesthetic quality).

This is one of the scientific injection molding methodology's most important studies or tools. However, for the final selection of the optimal injection speed – taking into account the aforementioned machine and mold limitations – we need to carry out some other studies or use other tools that will be explained in the following.

7.2.1.4 Influence of Melt Temperature on the Study

The temperature of the molten material obviously affects the viscosity of the material. This is an inversely proportional relationship (i. e. the higher the temperature, the lower the viscosity, and vice versa). However, in the relative viscosity test or viscosity curve, we are looking for the range of speeds at which the viscosity remains constant without affecting the shear or injection speed. Therefore, carrying out this test at different melt temperatures will not alter the range of speeds at which the viscosity is most constant. The viscosity of the material may vary, but the speeds at which the viscosity is most constant do not vary. We can therefore determine that the temperature at which the test is carried out is not important to the result we are looking for.

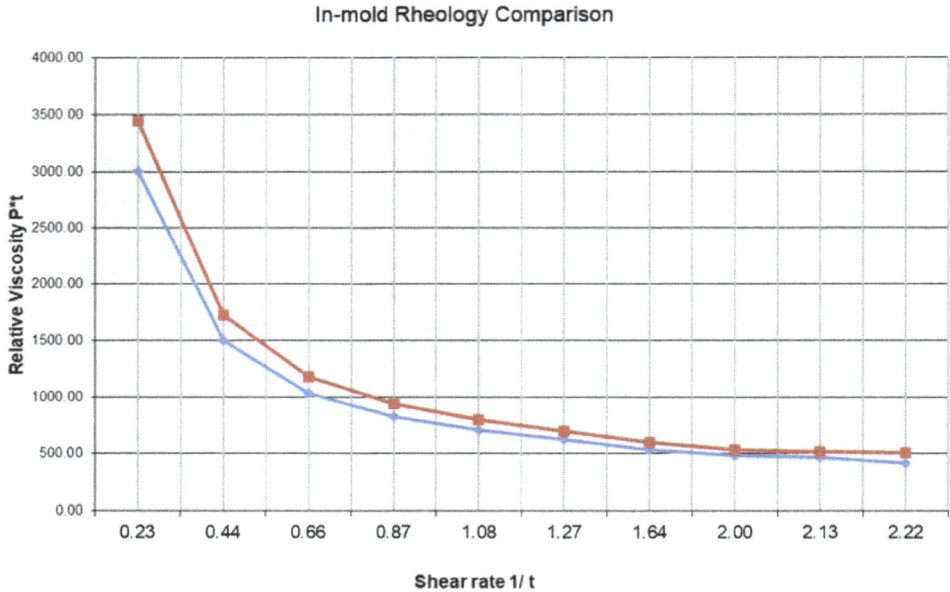

Figure 7.13 Viscosity curves at different melt temperature (blue curve at higher melt temperature than red one)

7.2.2 Injection-Speed Linearity Test

If we look at the table of data collected and calculated in Table 7.2, we can see that the viscosity decreases 5.5 times (between the maximum and minimum viscosity) just by varying the injection speeds in the range of the tests carried out. If the most important factor in determining the viscosity of the material is the shear rate and this is controlled by the mold filling speed, this means that we must control the injection speed to keep the shear rate as repeatable and constant as possible. Our machine must repeat the injection speed in a precise, repeatable way in order for us to have a robust process. We can control this through the output that will indicate if this injection speed is repeatable: the filling time. The logical conclusion would be that a constant, repeatable melt flow speed equals a constant shear rate, a constant viscosity, constant properties and a robust, consistent process.

One of the machine limitations with respect to the injection speed is the machine's ability to reach and maintain the setting's required injection speed. Usually, at higher injection speeds, the machine could not reach them quickly. So, we can check the injection machine's performance level across all the available injection speed settings.

During volumetric filling of the mold, we already define some of the molded parts' characteristics. From the perspective of the injection molding process, cavity filling is a speed-controlled setting. The injection machine must meet the speed setting with maximum precision which allows us to precisely control the speed at which the polymer fills the cavities of the mold.

The injection time, which is the output related to the injection speed, is the most critical time during the entire injection molding process. The injection molding machine must meet the speed setting at all available speeds, from slower speeds to the maximum speeds that the machine can reach. It must also rigorously repeat the filling time. Dispersion in the filling times means dispersion in the characteristics of the molded parts, and therefore not a very robust and consistent process.

In order to check the injection molding machine's ability in terms of the injection speeds, we can carry out an injection-speed linearity test, which basically consists in carrying out a sequence of injection shots at different speed settings, from slow to fast speeds from the injection machine range. So, if we know the injection stroke of the screw and the injection speed setting, we can calculate the theoretical time that the machine must spend to cover the injection stroke (i. e. we can arithmetically calculate the injection time); see Chapter 3. For example, if the injection machine has a net injection stroke setting (from the metering screw position, metering stroke + decompression stroke) to the switchover point to holding pressure of 100 mm and the injection speed setting at 100 mm/s, the screw must perform the injection shot in 1 second (i. e. the injection or filling time must be 1 second). If we set the injection speed (with the same net stroke of 100 mm) to 50 mm/s, then the injection or filling time must be 2 seconds. By performing the sequence of injection speeds in the whole range of injection speeds, we will be able to evaluate the response of our injection machine for the whole range of injection speed settings.

Injection-Speed Linearity Test: Injection Machine Configuration

1. Set the melt temperature in the machine and check that the melt temperature is the one recommended by the raw material manufacturer.

2. Set the machine with a holding pressure and holding pressure time of zero.

3. Set a high injection pressure limit so that it neither affects nor limits the process.

4. Set a cooling time to ensure that the parts are solid when demolded.

5. Set the switching (from filing to holding) system by stroke or volume.

6. Ensure that the switchover point is somewhat premature as it could cause flash and damage the mold when testing at high speeds (it could be 95% of the total volume, for example).

7. Set the lower test injection speed.

8. Proceed with a regular increase in the injection speed setting – the objective is to test and carry out shots at at least 10 different speeds between the minimum and maximum available machine injection speed range.

9. In each cycle at different injection speeds, we must collect and record the data corresponding to the maximum injection pressure at the switchover point and mold filling time. With these data, we will prepare the table for the calculation of the real injection speed reached and the machine injection speed level achieved.

Table 7.3 Information Needed for the Injection-Speed Linearity Test

Metering stroke	55 mm
Switchover point	33 mm
Net injection stroke	22 mm

Net injection stroke = metering stroke − switchover point

Table 7.4 Example of a Sequence of Collected Study Data

Shot Number	Required Speed (mm/s)	Filling Time (s)	Filling Pressure (bar)	Real Speed Obtained (mm/s)
1	5	4.4	782	5.00
2	10	2.25	766	9.78
3	15	1.51	782	14.57
4	20	1.15	814	19.13
5	25	0.93	853	23.66
6	30	0.79	884	27.85
7	40	0.61	974	36.07
8	50	0.5	1052	44.00
9	55	0.47	1089	46.81
10	60	0.45	1115	48.89

In the resulting graphs (Figure 7.14), we can see the blue line that corresponds to the set speed and the red line that corresponds to the real speed reached. The distance between the two lines indicates the level of "non-compliance" of the injection speed. The greater the gap between these two lines, the greater the level of "non-compliance" between the set speed and the real speed reached.

Figure 7.14 A graph showing the injection speed from the injection-speed linearity test

Normally, as we increase the speed – as in this example – the level of injection speed mismatch is greater. This is because the screw needs time to accelerate to reach the set speed; this means that the speed setting of the full injection stroke cannot be reached.

If we carry out this study on many occasions and compare the level of mismatch, over time we will be able to detect problems concerning the calibration or linearity of the machine, acting scientifically to detect deviations in the machine injection speed ahead of time.

It is acceptable to have a minimum speed compliance level, at high injection speeds of 85% to 95%. That means an injection speed non-compliance of 10% to 15%.

For the repeatability of the injection process, it is essential that the injection speed be repeatable. We cannot repeat a process with an injection molding machine that has variability in its real injection speed, either due to wear or decalibration.

If we do not apply this study frequently, we will have spent time, effort and resources on defining parameters to approve a process and its parts and, after a while, our parameters will not be valid unless we have the machine under control from an injection-speed linearity standpoint.

Using the "Injection Speed Machine Performance" spreadsheet (Figure 7.15), we can calculate and plot the level of compliance of the injection speeds set on the machine control unit and whether the injection speeds are properly calibrated. By carrying out the linearity speed study, we can check the machine's precision, given that the injection speed is the most critical speed in the entire process.

SCIENTIFIC MOLDING   CALCULATIONS AND PORTABILITY

LINEARITY INJECTION SPEED

		mm
DOSAGE STROKE	55	mm
SWITCHOVER POINT	33	mm
FILLING NET STROKE	22	mm

Put data in green cells   Menu

INJECTION SPEED	FILLING TIME	INJECTION PRESSURE	REAL INJECTION SPEED RESULTS	SPEED PERFOMANCE
mm / sc	sc	bar	mm/sc	%
5	4.4	782	5.00	100.00
10	2.25	766	9.78	97.78
15	1.51	782	14.57	97.13
20	1.15	814	19.13	95.65
25	0.93	853	23.66	94.62
30	0.79	884	27.85	92.83
40	0.61	974	36.07	90.16
50	0.5	1052	44.00	88.00
55	0.47	1089	46.81	85.11
60	0.45	1115	48.89	81.48

Air Shot Test

35	0.69	200	31.88	91.10

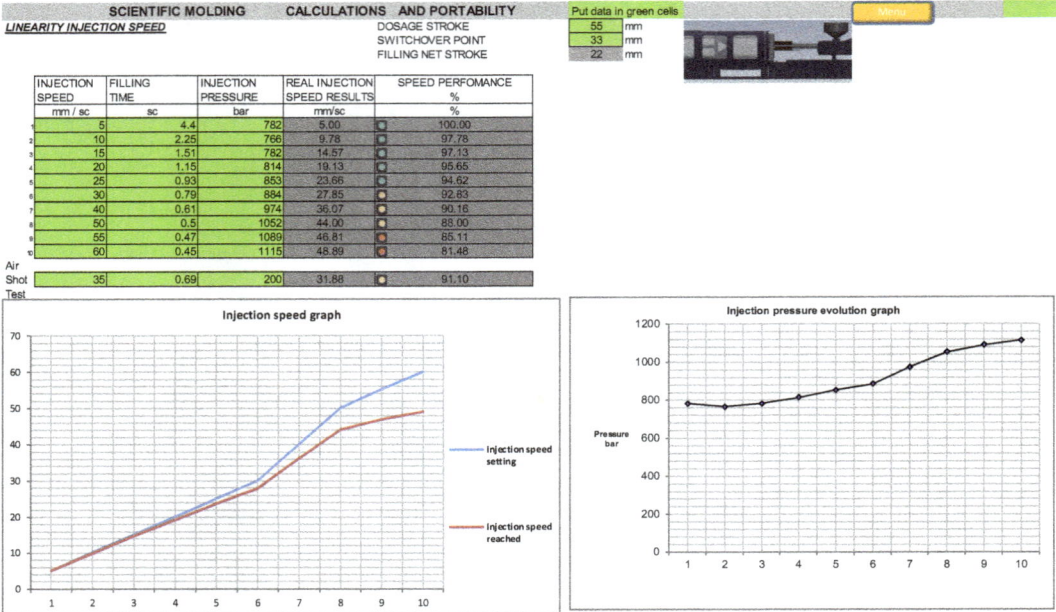

Figure 7.15 Example of a spreadsheet for performing the study (by entering the values in green cells, we can obtain the calculated values and linearity speed graph)

7.2.3 Mold Filling Balance Study

As the molten material fills the mold, it begins to cool and flow into the cavities at a given speed, decreasing temperature and pressure. Filling a mold is a "time-dependent" process. This is because, depending on the time factor, the other variables involved in filling the cavity will also vary. These variables are:

- Melt temperature
- Injection pressure
- Material flow speed

All of them are time-dependent; in the short period of time during which filling occurs, these variables will change significantly and affect the properties of the molded parts, dimensions, aesthetics, etc. The quality of the molded parts depends on the uniformity of these variables in all the mold cavities. All this means that, in order to fill multiple cavities under the same pressure, temperature and speed conditions, it is necessary to perform correct, balanced filling of the cavities. The goal must be to reach the end of the filling stage, with the same conditions in all the different cavities, resulting in the same part properties and consistency in the process.

What might be considered acceptable levels of cavity imbalance? This depends on the type of part, the sector and the precision required – as a standard value, we can consider levels below 3% imbalance as "correct" but this should be tested to determine the maximum permissible in cases of tight tolerances and multi-cavity molds.

7.2.3.1 Filling Balance Study: Injection Machine Configuration

1. Set the holding pressure and holding pressure time to zero.

2. Set the cooling time to ensure correct demolding.

3. Set the injection speed to check the filling balance.

4. Set the switchover point for the holding pressure to reach a cavity fill of at least 95–98% of the part volume.

Note that, in the case of low viscosity material or soft material, it may be necessary to set a delay time before starting the metering, so as to avoid material overfilling the mold during metering, due to back pressure.

With parts filled just to the switchover point, we can weigh them and can calculate the filling balance of the mold. Deviations up to 3% are considered acceptable.

It is possible to check with different injection speeds (slow, medium, and fast), since the different speeds, with cold distribution channels, can modify the filling balance. The following injection speeds can be selected for the test:

- High injection speed – maximum injection speed obtained from the injection-speed linearity study with an injection speed non-compliance of 10% to 15%

- Low injection speed – minimum injection speed obtained from the viscosity curve study, where the Newtonian behavior of the polymer starts, in the most stable and flat portion of the curve (near the "elbow")

- Medium injection speed – average injection speed of high and low injection speed selected previously

Calculate the weight averages of 10 shots in a row to determine the deviations. Consider the possible spiral effect mentioned in the next section.

Table 7.5 Sample Filling Balance Study Conditions (with Speeds Selected for Testing)

	High Speed	Medium Speed	Low Speed
Speed (mm/s)	60	30	20
Injection time (s)	0.5	1	1.5

Table 7.6 Sample Data Collected from a Filling Balance Study in an Eight-Cavity Mold

Cavity	Average Weight over 10 Injection Shots (g)		
	High Speed	**Medium Speed**	**Low Speed**
1	1.27	1.31	1.30
2	1.27	1.30	1.30
3	1.27	1.32	1.31
4	1.30	1.30	1.30
5	1.28	1.30	1.30
6	1.31	1.31	1.30
7	1.32	1.30	1.30
8	1.30	1.30	1.30

Table 7.7 Filling Balance Calculations at Different Injection Speeds

Calculations	High Speed	Medium Speed	Low Speed
Max. weight (g)	1.32	1.32	1.31
Min. weight (g)	1.27	1.3	1.3
Range (g)	0.05	0.02	0.01
Imbalance (%)	3.8	1.5	0.8

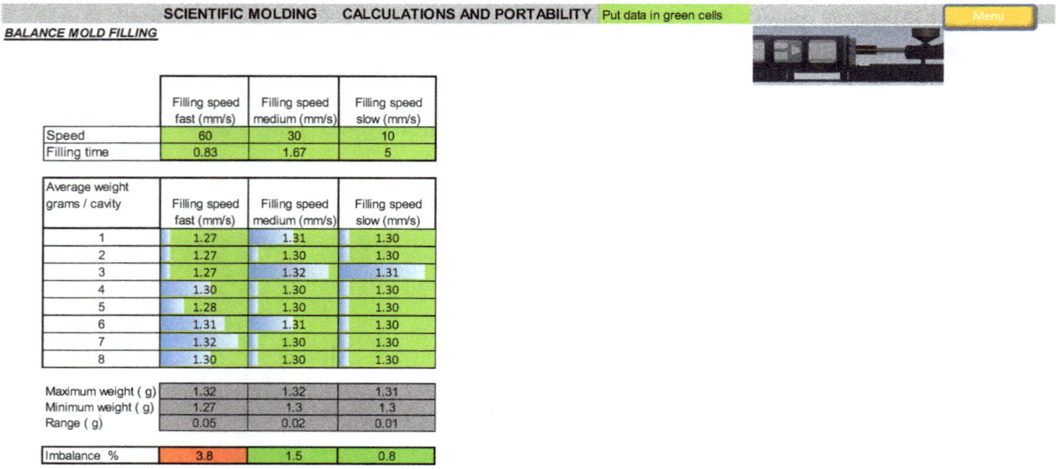

Figure 7.16 Example of a spreadsheet for performing the study (by entering the values in the green cells, we can obtain the calculated values of mold-filling balance)

7.2.3.2 Spiral Effect

How many times have you seen a multi-cavity mold with a fully symmetrical and balanced runner that fills the cavities in an imbalanced manner? (i. e. while some cavities are filled, others are only partially filled). In that event, we suspect that the root cause of the filling imbalance is that the cavity gates are not identical. However, after checking, all the gates all have the same dimensions. So, what could be happening to cause the filling imbalance?

It is most likely the spiral effect or shear-induced mold-filling imbalance. This undesirable effect is caused by the influence of several factors. It all starts with the material feed system inside the screw, so let's take a look.

InFigure 7.17, we can see that at the end of the screw, the material is undergoing rotational cr mixing flow, which is promoted by the rotary motion of the screw during metering, approximately up to the machine nozzle. From this position, however, the type of motion changes – it becomes laminar flow.

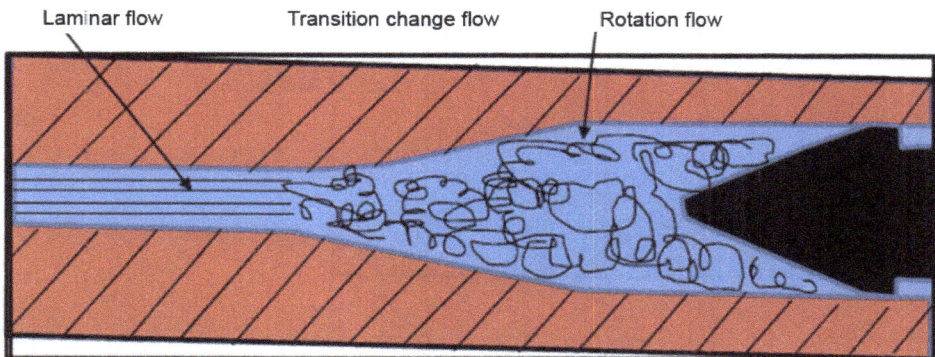

Figure 7.17 Spiral flow origin diagram

Starting from the transition zone, the material will advance in laminar flow, as parallel layers. Incidentally, if a defect (such as black spots or degradation) always occurs in the same position on the part, we must suspect elements that are located behind the transition line (i. e. nozzle, nozzle holder, hot runner, etc.). On the other hand, if a defect occurs in random areas of the part, we must suspect elements in front of the transition line (i. e. screw tip, screw fillet, etc.).

Therefore, laminar flow of the material is one of the factors that cause the spiral effect. Another factor is the viscosity of the material, since the spiral effect appears when molding materials of medium or high viscosity. The third and final factor is a multi-cavity mold with more than four cavities.

As the molten material fills the mold, the material in contact with the mold steel cools rapidly, creating a cold skin. The layer immediately below, circulating within the molten material, is subject to friction or shear between the central flow in motion and the static and solidified cold layer. The maximum shear is generated in this boundary layer and the minimum shear is generated in the center of the flow. This friction causes the layer immediately below the cold layer to reach a slightly higher temperature than the central flow area. In Figure 7.18, red represents hotter material, due to friction, and blue represents colder material.

These layers move in laminar flow, with the central layer moving faster, but the zone between this central layer and the outer cold layer experiences the highest level of shear. This condition of shear between layers is called "shear thinning", in other words, a reduction in viscosity, due to high shear or stress between layers. This effect occurs particularly in cold channels, but also in hot runners channels.

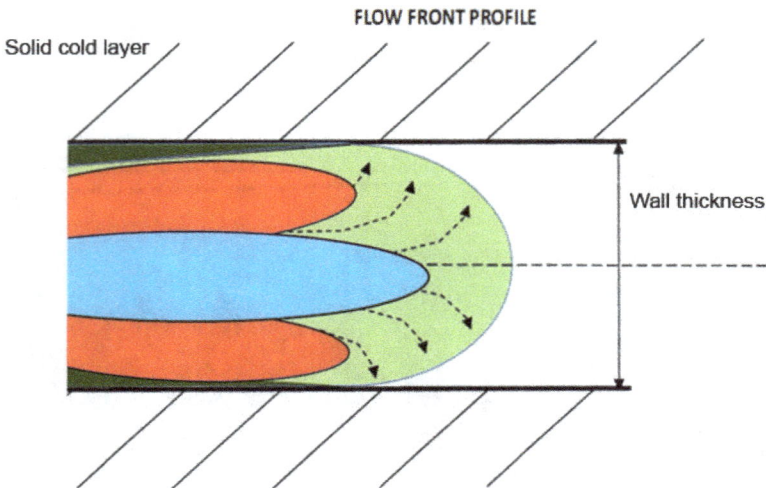

Figure 7.18 Different temperature layers during filling

Figure 7.19 Differences in cavity filling, due to spiral effect

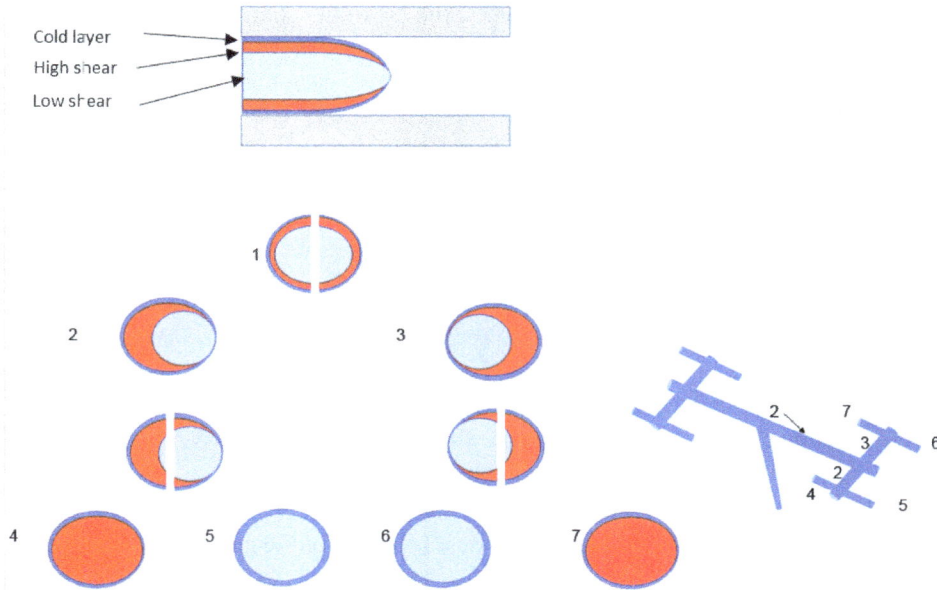

Figure 7.20 Separation of the hottest and coldest flows

This effect can be minimized with a patented flow modification system called Melt-Flipper®, which has been available for some time. It consists of placing these elements in the distribution runners so that the hotter flow mixes with the colder flow, homogenizing the cavity filling. MeltFlipper® – find more information at: *www.beaumontinc. com/meltflipper/*.

In an eight-cavity mold, there are two flows of material with different filling behavior – one flow is for filling the four cavities near the center of the mold (the one that will fill faster) and the second is for filling four cavities further outside the channel (the one that will fill more slowly). In the case of 18 cavities, there will be four different flows; in the case of 32 cavities, there will be eight different flows.

It is advisable to analyze the possible filling problems (properties, dimensions, etc.) as a function of the cavities that belong to each flow or filling behavior.

7.2.4 Optimal Injection Speed Selection

With the studies discussed in the previous sections:

■ Relative viscosity method or in-mold rheology

■ Injection-speed linearity

■ Filling balance

we have important information for selecting the most robust and consistent mold fill-ing speed, and therefore the most suitable for an optimal injection process. These three tests will help us determine the injection speed at which the process will be most robust and consistent.

Finally, we will add another important factor in the quality of the molded parts to these three studies: **aesthetic quality**. This aesthetic factor is important when select-ing the injection speed.

Relative viscosity curve or in-mold rheology:

In this study, we have constructed the viscosity curve of the material as a function of the shear rate applied, and we have seen that it is L-shaped or elbow-shaped. We can also see that, at a low shear rate and low injection speed, the viscosity is high and af-fected more by the shear rate and that, when the injection speed increases and passes the elbow zone of the curve, the material exhibits Newtonian behavior, its viscosity is more constant and is not affected by changes in the injection speed.

From the resulting viscosity curve, we know the minimum speed at which we should inject to avoid non-Newtonian and non-robust behavior of the material. We know the speed range over which we have a process whose viscosity is heavily influenced by the injection speed. The speed that will be selected will allow us to inject with a robust process that will have a constant and repeatable viscosity because small changes in speed will not affect the viscosity in this portion of the curve, and neither will changes in viscosity affect the speed.

We can draw the minimum speed line to be used. This speed should not have relative viscosity in the vertical portion of the curve and should be a speed that will result in constant viscosity in the flat portion of the curve (Figure 7.21).

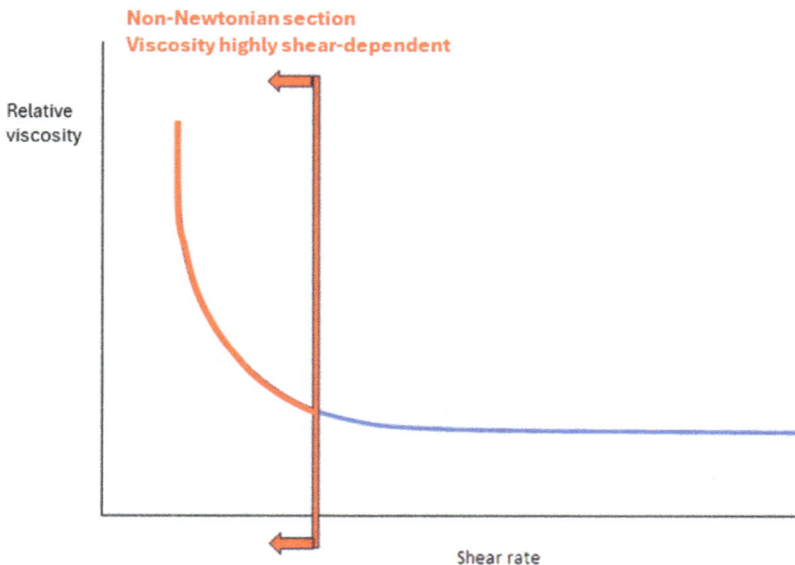

Figure 7.21 Section of the viscosity curve exhibiting non-Newtonian material behavior

Injection-speed linearity:

If speeds within this flat, pseudo-Newtonian portion are the most suitable for filling the mold, does this mean that the most suitable speed will be the machine's maximum injection speed? At high speeds, undesirable effects can occur that otherwise harm the process and some of the process limitations can appear, including mold issues, venting, burrs, flash, not forgetting that we must consider that high injection speeds can damage the plastic material and/or its additives, due to excessive shearing.

With the aid of the injection-speed linearity test, we can see that when the injection speed is increased, the level of mismatch between the set speed and the real speed is greater. This is because the screw needs time to accelerate to reach the set speed, so it cannot reach the set speed for the entire injection stroke; the greater the distance between these two lines, the higher the level of non-compliance between the set speed and the real speed reached.

At high injection speeds, a minimum speed compliance of 85–90% is acceptable. In other words, a maximum level of non-compliance of 10–15% is acceptable (Figure 7.22). So it is speeds in this flat, pseudo-Newtonian portion that we are interested in using for our mold filling, but does this mean that the most suitable speed is the machine's maximum injection speed?

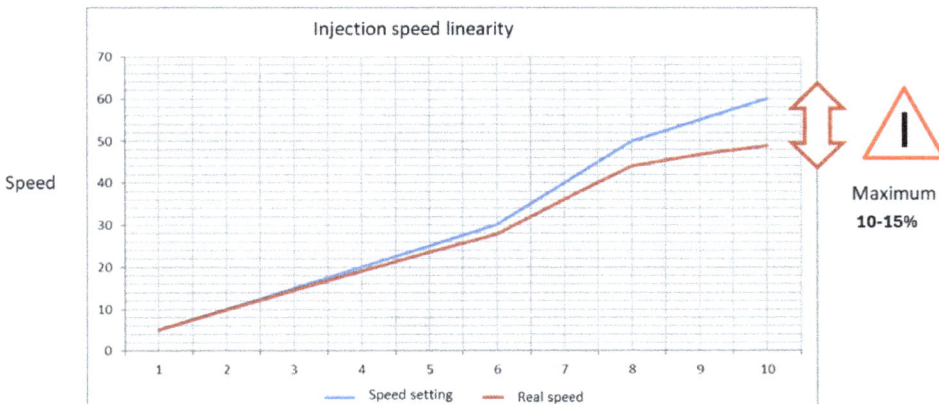

Figure 7.22 Graph for the linearity test

With the data from the linearity test, we can define the maximum speed within the material's Newtonian behavior range and an injection speed at which the machine can reach at least 85–90% of the setting (Figure 7.23). We can then draw the recommended maximum speed line.

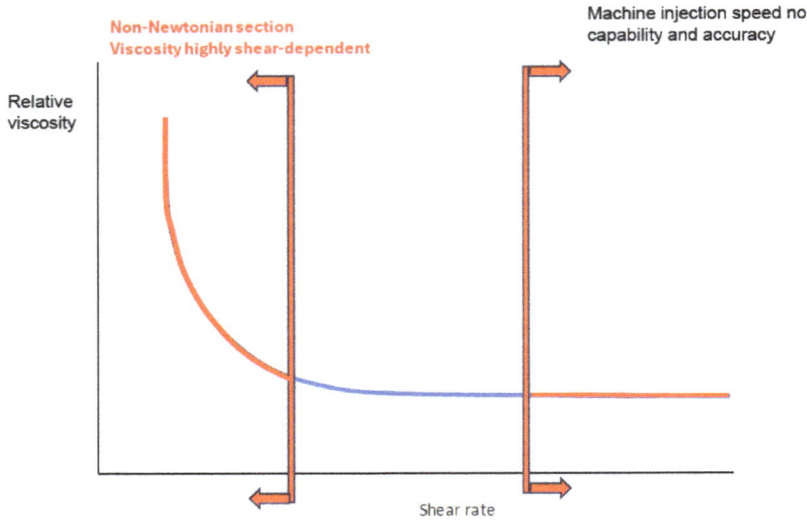

Figure 7.23 Portion of the viscosity curve, with uncontrolled injection speed and unattained speed

Filling balance:

This is the third test that will help us select the most robust and consistent injection speed. Section 7.2.3 explains how to test the mold-filling balance with three injection speeds selected from the viscosity curve. These will be: low speed, the minimum injection speed of the viscosity curve, where the Newtonian behavior of the material begins; high speed, the maximum injection speed from the injection-speed linearity test (i. e. where a maximum mismatch of 10–15% is reached); and the medium speed, the average speed of the other two speeds mentioned (Figure 7.24).

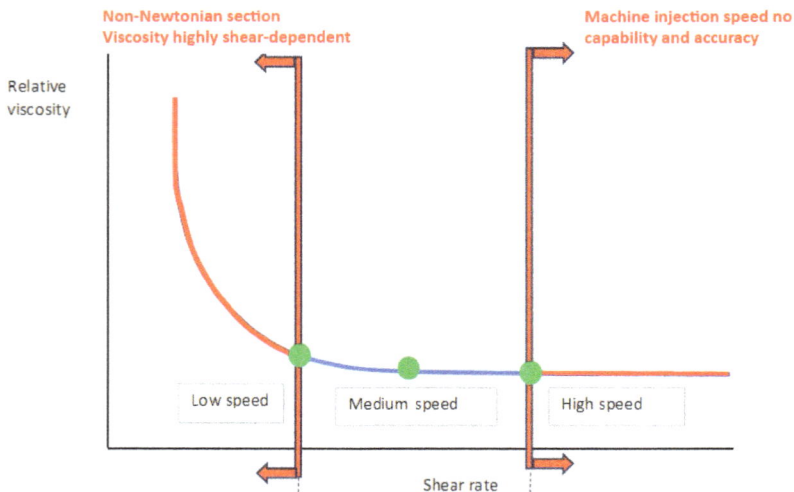

Figure 7.24 Three selected injection speeds for performing the filling balance test (low, medium, and high)

SCIENTIFIC MOLDING   CALCULATIONS AND PORTABILITY Put data in green cells

BALANCE MOLD FILLING

	Filling speed fast (mm/s)	Filling speed medium (mm/s)	Filling speed slow (mm/s)
Speed	60	30	10
Filling time	0.83	1.67	5

Average weight grams / cavity	Filling speed fast (mm/s)	Filling speed medium (mm/s)	Filling speed slow (mm/s)
1	1.27	1.31	1.30
2	1.27	1.30	1.30
3	1.27	1.32	1.31
4	1.30	1.30	1.30
5	1.28	1.30	1.30
6	1.31	1.31	1.30
7	1.32	1.30	1.30
8	1.30	1.30	1.30

Maximum weight (g)	1.32	1.32	1.31
Minimum weight (g)	1.27	1.3	1.3
Range (g)	0.05	0.02	0.01

Imbalance %	3.8	1.5	0.8

Figure 7.25 Example of a filling balance study (in this case, the most balanced speed is the slower speed)

Aesthetic quality:

In many cases, the aesthetic aspects of the molded parts are critical for their production. It is therefore necessary to take aesthetic considerations into account when selecting the most suitable injection speeds for the process. Sometimes this factor forces us to use injection speeds that are not the most suitable.

Injection Speed Selection

Using the information from the studies or tools explained in this section, we can select the most suitable injection speed from the injection speed window. That is to say, considering the relative viscosity method or in-mold rheology, the injection-speed linearity, the filling balance, and the aesthetic concerns, we make a scientific selection of the injection speed and select an injection speed from the final process window (Figure 7.26). This selection will have accounted for variables such as:

- Material: viscous behavior of the material as a function of shear rate or injection speed
- Mold: filling balance
- Machine: the injection molding machine's ability to comply with the speed setting
- Part: aesthetic results

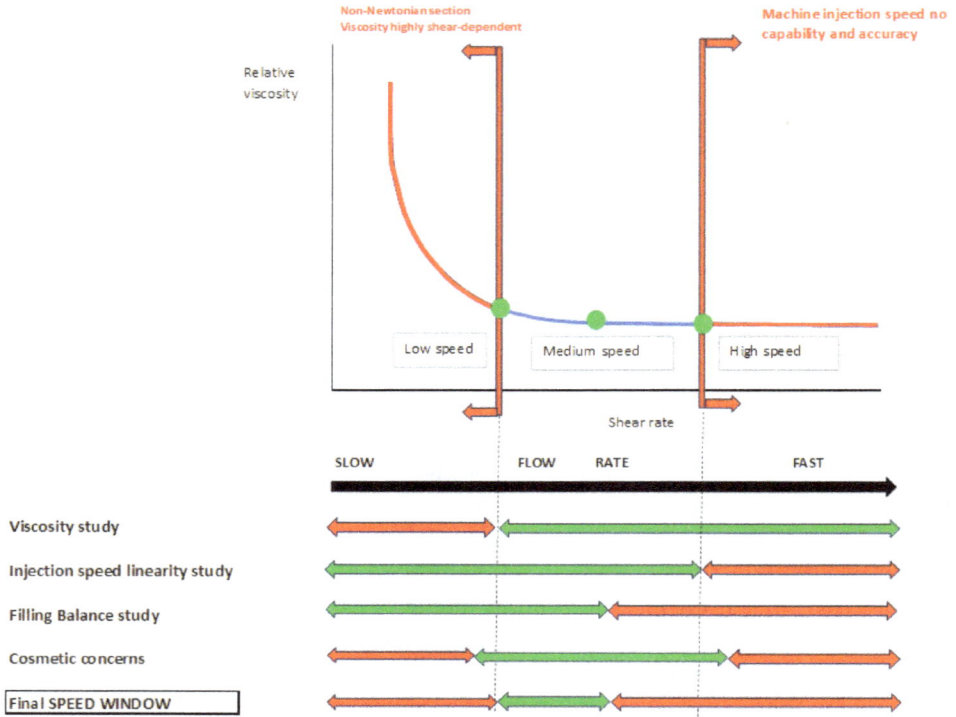

Figure 7.26 Injection speed selection

7.2.5 Mold Filling Study

Within the trials, studies and tests that should be carried out with a new mold, one of the most important studies is to analyze how the molten material fills the cavities of the mold. To do this, it is necessary to conduct the mold filling study, which is an essential tool of the scientific injection molding methodology.

Why conduct the mold filling study methodology?

1. For the safety of people and the mold

2. To avoid possible overfilling, excessive cavity pressure, hold, etc. during startup and testing

3. To properly determine the right switchover point from the filling- to holding-pressure stage

4. To properly set the switchover point (injection pressure, screw position, filling time)

5. To collect information that will allow us to later select the most suitable switching system – switching by filling time, by injection pressure, by screw position or volume, or by cavity pressure

6. To evaluate the precision and repeatability of the volumetric filling during the filling stage, and repeatability of the switchover point

7. To evaluate the filling balance of the different mold cavities

8. To check for a possible spiral effect during filling

9. To evaluate the root causes of certain defects in the molded parts, such as shrinkage, burrs, weld lines, bubbles, shine, flash, streaks, etc.

10. To evaluate the differences between material batches in terms of viscosity and rheology

11. To evaluate the pressure losses in the flow path (explained in Section 7.2.6)

12. To evaluate the sealing of the screw check ring valve

13. Finally, to determine exactly where the melt flow accelerates or decelerates in order to define the points of change in the injection speed profile, if required

Be careful! Before carrying out a mold filling study, it is important to check if there could be a demolding issue when molding very incomplete parts. In some molds, when very short parts are filled, ejection of the incomplete part may be complicated and may even cause some damage to the mold, or in other cases, it may be necessary to spend a long time removing the short-molded parts. In such cases, we should not carry out the mold filling study.

7.2.5.1 Mold Filling Study: Injection Machine Configuration

Progressive mold filling settings (Figure 7.27):

- Melt temperature as recommended

- Mold temperature as recommended

- Set a high injection pressure limit so that the process cannot be pressure limited

- Set the holding pressure values to zero (another option is to set enough holding pressure to prevent screw rebound)

- Set the holding pressure time to zero (another option is to set enough holding-pressure time to prevent screw rebound)

- Select the switching system based on the screw position or volume

- Set a switchover point to holding pressure close to the position of the screw metering stroke

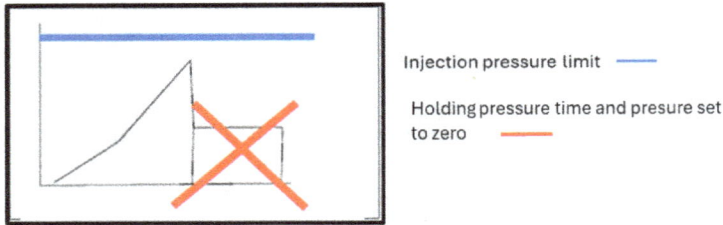

Injection pressure limit ——

Holding pressure time and presure set to zero ——

Figure 7.27 Machine configuration for the mold filling study

7.2.5.2 Mold-Filling Study Procedure

The study should be done gradually, changing the switchover point with each shot. Thus, each shot will increase the volume of material placed into the mold.

1. Fill the mold progressively, checking how the runners and cavity fill by gradually moving the switchover point. The goal is to gradually increase the injected volume step by step.

2. Pay special attention to the points where the flow accelerates and decelerates. Observe the screw position of the injection stroke at which the acceleration and deceleration of flow occur.

3. If necessary, change the injection speed at the screw positions where the material flow front accelerates and decelerates. The goal is to obtain a uniformly advancing flow front, due to the shear thinning behavior of the polymer as mentioned in Section 4.3.2.

4. Determine the optimum switchover point for the holding pressure stage. Volumetrically fill the cavity to 95–98% (under normal conditions, the material will have reached every corner of the cavity).

   At this point, the machine control provides the following information:
   - Filling time: Use this data for setting if you want to switch based on time
   - Injection pressure at switchover point: Use this data for setting if you want to switch based on injection pressure
   - Real switchover point position or volume: Use this setting if you want to switch based on stroke or volume

It is advisable to retain samples of each shot made and to use an indelible marker to mark the shot with the values of the switchover position, filling pressure and injection time; see Figure 7.28. These marked samples show us the exact filling of the mold and we can easily see the points where the filling speed changes and where the flow front accelerates or decelerates (sometimes even where it stops).

These filling-speed switchover points, based on the marked parts injected during the progressive filling test, are the real change points – they are exactly where the flow front changes speed. For example, if the flow front creates a weld line during the mold filling study, we can know exactly where in the mold filling path and screw position the material is passing through that critical area. We can therefore specifically modify the injection speed at the position where the defect is formed.

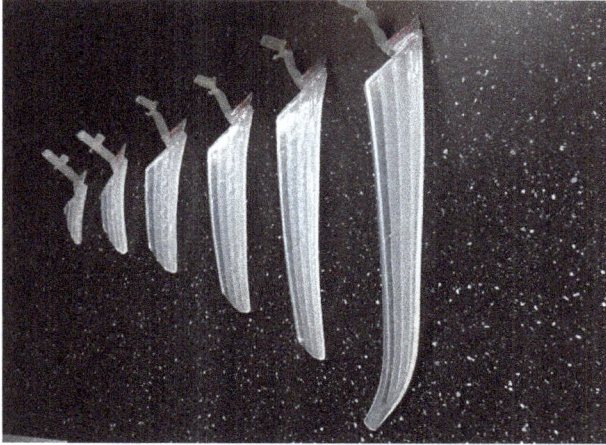

Figure 7.28
Example of parts from a mold filling study

Remarks:

- Sometimes (especially with low-viscosity materials) during the metering phase and due to back pressure, the material can be pushed and fills the mold. This material that fills the mold during metering will alter the volume that fills the cavity during the filling phase and will affect the evaluation of the cavity filling. This effect can be avoided by setting a delay between the filling and the metering phases, usually called "metering delay time".

- We must always have a screw front cushion available throughout the mold filling study.

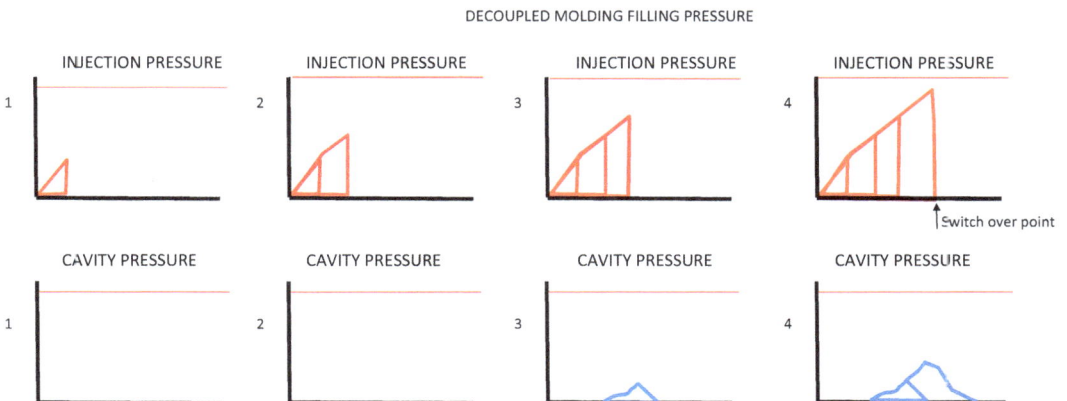

Figure 7.29 Development of injection pressure during the mold filling study

By continuing the mold filling study and obtaining partial injections, we will be able to observe where the flow front accelerates, decelerates or even stops, and this will allow us to configure an injection speed adapted to the reality of the flow progress.

Figure 7.30 Development of short-shot parts during the mold filling study

Switchover point:

By carrying out the mold filling study in the final phase of filling, we can determine the optimal switchover point. Filling 95–98% of the cavity volume in dynamic filling is recommended, so we can decide what percentage of cavity filling to reach before changing to the packing and holding phase. Collect and record the switchover control data:

- Injection filling pressure

- Filling time

- Screw position at switchover point

- Part weight at switchover point

With these output data, we can choose which switchover system to set. Once we have selected the switchover point, we can proceed to check the precision of the machine at this point. To do this, we carry out several shots and weigh the recently filled parts (including the cold runner) to check the stability and consistency of the shot weight. At this point, theoretically, if we change the switchover system based on screw stroke, injection pressure, or filling time, we should obtain approximately the same part weight regardless of the switching system selected.

A final recommendation is that, once the switchover point and the switching system have been selected, the weight of the parts just filled at the switchover point and un-packed should be recorded and monitored as an important process output.

7.2.6 Analysis of Injection Pressure Losses along the Filling System

During filling, the plastic in contact with the mold begins to cool, forming a cool skin due to the mold's cooling system. The thickness of this skin increases on account of such cooling. Furthermore, the boundary layers of the flow in contact with the cool skin generate shear stresses as they advance during filling, which in turn generates heat. This pseudo-balance of cooling and heating allows the filling of large cavities and long flow paths. However, the pressure required to push and advance the flow front through distribution channels, sprues, gates, and cavities increases, generating an inevitable loss of pressure at each step on the flow path.

The injection molding process is a deficit process from an injection pressure stand-point. If we compare the injection pressure generated by the hydraulic pump – in the case of hydraulic machines – with the pressure in the injection piston and the pressure in the sprue, runner and cavity, we can observe the pressure losses generated by the restrictions, thickness, etc. in the flow of the material and its progressive cooling.

The injection molding machine requires a certain level of injection pressure to achieve the programmed injection speed. This required total injection pressure should never exceed the maximum programmed pressure or the injection pressure limit. If exceeded, this would be a pressure-limited process, which should be avoided because it is a situation out of the process's control.

By knowing the pressure losses in each process constraint, we can assess whether measures must be taken to reduce any pressure losses and achieve a normal total pressure loss for the injection pressure required to fill the cavities, ensuring that we never reach the maximum injection pressure limit of the machine and that there is enough pressure to have a suitable Delta P.

To analyze the pressure loss generated by each of the steps in the filling path system, from the screw tip to the cavity, it is advisable to carry out a pressure loss analysis. It is also advisable to take advantage of the development of the progressive filling test to determine the pressure of passage of the flow through the sprue, the main runner, the secondary or sub-runners, the gates, the cavities and, of course, the pressure required to eject the material into the air through the machine nozzle.

From these conveniently presented data, we can see the percentages of pressure loss and pressure drops for each step on the material's filling path from screw tip to cavity. This analysis enables us to make decisions and prioritize them to reduce the pressure losses in the system, if necessary. **Remember that the higher the pressure losses, the greater variability of the process when material viscosity changes**.

With this tool, we can check how much pressure is required to reach certain positions in the filling path or flow path of the molten material. These are:

- Pressure required for purging

- Pressure required to fill the sprue

- Pressure required to fill the runner to the gate (without passing through the gate) – here we can have main, secondary, and tertiary channels

- Pressure required to pass through the gate

- Pressure required to fill the cavity

In the case of hot runners:

- Pressure required to pass through the hot runner only

- Pressure required to pass through the possible cold runner and gate, if any

- Pressure required to fill the cavity

Pressure Loss Study: Injection Machine Configuration

- Set the injection pressure limit to the maximum limit allowed by the injection machine.

- Set the holding pressure time and holding pressure to zero.

- Set the mold filling speed, following the viscosity curve study, injection-speed linearity test and filling balance study, and injection speed selection; Section 7.2.4.

- Purge or eject the material through the nozzle into the air and record the pressure reached.

- Carry out a progressive filling, as described in Section 7.2.5 ("Mold Filling Study"): by changing the switchover point to holding pressure, we fill only the sprue, then the main runner, sub-runners, the cavity gates, and finally, the cavities. Record the maximum pressure needed to reach each position.

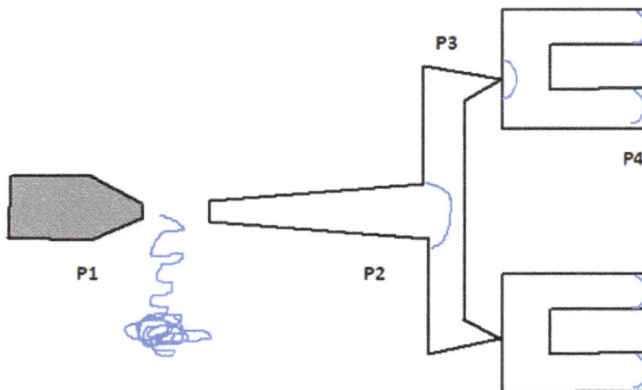

Figure 7.31 An example of filling positions for a pressure loss study

Figure 7.32 shows a sample spreadsheet for studying pressure loss along the flow path into the major pressure consumers for further analysis. If any of them are excessive, we can modify thicknesses or the design to reduce this.

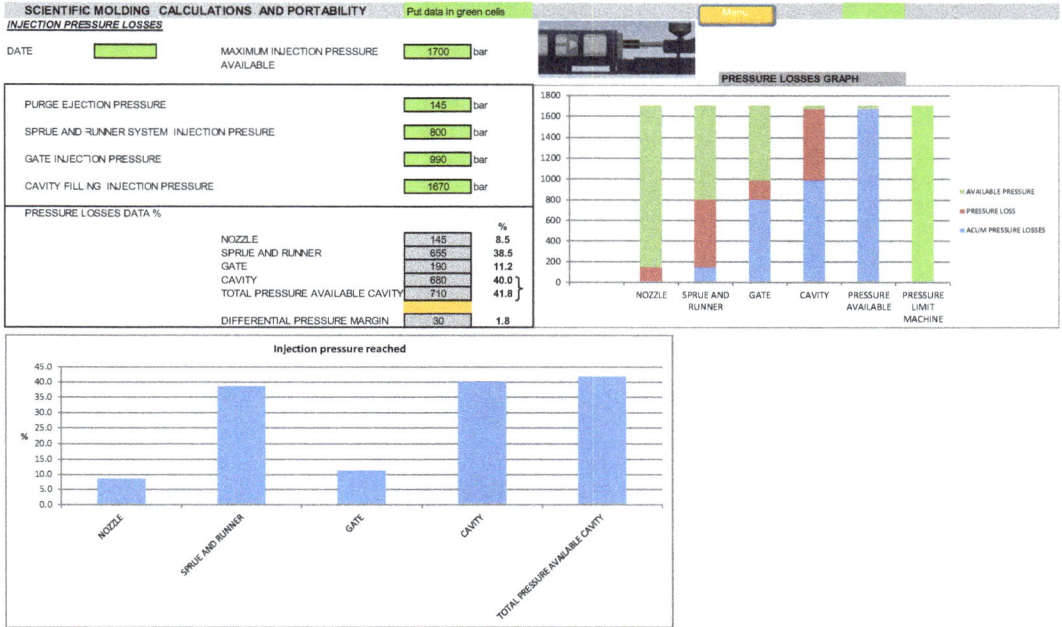

Figure 7.32 Example of a spreadsheet for a pressure loss study

In Figure 7.32 and Figure 7.33, we can see that at the end of the cavity filling, only 1.8% of the total machine pressure is available for possible pressure readjustment. We are very close to having a "pressure-limited process". To avoid this problem in the example, it would be necessary to improve and reduce the pressure drop in the sprue and the runners, which is high, namely 38.5%. It is recommended that the total maximum pressure used to fill the runner system and the cavities be 90% of the maximum available pressure in the injection machine.

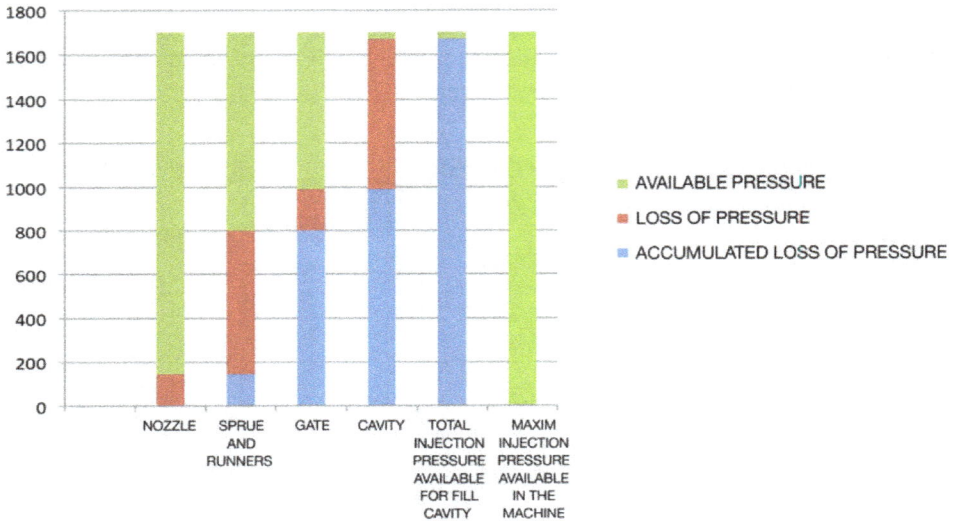

Figure 7.33 Example of a graph of pressure loss along the flow path

7.2.7 Delta P Determination and Study

By setting the right Delta P value, the process can adapt the required injection pressure to the variability of the material viscosity. Note that the Delta P study included in Section 3.3 is used to check the machine performance and behavior regarding this important pressure. In this chapter, the focus is to set the Delta P pressure by following the scientific injection molding methodology.

Looking at the in-mold rheology test, Section 7.2.1, we can see the strong influence of the injection speed and shear rate on changes in material viscosity. Therefore, the injection speed needs to be very well controlled. The injection speed (and consequent filling time) is the most critical speed and time of the process variables, so the filling time should have a very tight tolerance. Because of the strong influence of the injection speed on the polymer viscosity (shear thinning), some experts recommend tolerances of ±0.04 seconds for the filling time. Due to these tolerances, the machine, not the technician, must regulate itself to maintain constant filling-time cycles and consistent part quality. For this machine to self-regulate conditions for this critical parameter, an appropriate Delta P setting is mandatory.

As explained in Section 3.3, an analogy with the cruise control of a car can be drawn. If we set the cruising speed (e. g. 100 km/h) and need to climb a steep slope, the vehicle will require more power to maintain the set speed; on a steep descent, the vehicle will need to reduce its power because it is not necessary to reach the set speed. The same concept applies to the filling control in the first phase of volumetric cavity filling in an injection molding machine. However, in the car we always have 100% engine power available, while in our injection machine we can regulate this available "power", which is the injection pressure or the injection pressure limit parameter.

Furthermore, if we set the injection pressure limit (equivalent to the power of the car) to the maximum, and if one or more cavities happen to be closed, the injection machine could use all the available energy (pressure) to reach the set injection speed values and the switchover point to the packing and holding phase. This would cause damage to the mold and parts would be produced with defects, bad dimensions, flash, stress, etc. On the contrary, if we set an injection pressure limit lower than that needed to obtain the required injection speed, we will prevent the available machine pressure from reaching the set injection speed, giving rise to a "pressure-limited process". In the latter case, we would be slowing down the injection speed or filling speed, and it is strongly discouraged to inject parts using a pressure-limited process. This is a common mistake in some injection plants.

Delta P is the difference (in bar) between the injection pressure limit setting and the real injection pressure required to fill the cavities during the filling stage. We should not set the injection pressure to the maximum allowed unless this is necessary, nor should we cut the injection pressure curve. A poorly programmed Delta P may be lower or higher than the optimum setting:

- If Delta P is programmed with a value below the optimum: This is called a pressure-limited process. The machine will not be able to self-regulate the pressure required to reach the set injection speed. This situation should be avoided. It is one of the most serious mistakes in the injection process. When the required pressure reaches the set pressure limit, the curve flattens and the flow front slows down and loses control over cavity filling.

- If Delta P is too high and a cavity is clogged (e. g. in multi-cavity molds), there will be excessive flash and we may even damage the mold, parting lines, slides, etc.

A suitable Delta P value for standard injection molded parts is 15–30 hydraulic bar or 150–300 specific bar. The molder can reduce this Delta P if necessitated by part and mold requirements.

Since viscosity varies with different resin batches, colors, regrind material or moisture content, etc., we need the machine to automatically adjust to viscosity changes, so that we do not have to do it ourselves. If the machine has enough extra pressure (Delta P) and the new batch resin has a higher viscosity, the machine will use more pressure, like a car going up a hill, but the mold-filling time will be constant. If the material is of lower viscosity, the machine will use less pressure – like a car going down a slope – but the filling time will be constant. In fact, the machine must use a slightly different pressure for each cycle, due to the large number of variables that exist during the process.

The question now is how much extra pressure is required for the mold-filling time and adjustment for viscosity and other variations? 10–20% extra pressure is recommended, or the 150–300 specific bar already mentioned.

Delta P Determination

As discussed in Section 3.3, beforehand we should check the maximum injection pressure for each cycle during filling before switching to the holding pressure stage, and we should check the injection time.

Delta P Study: Injection Machine Configuration

1. The system of switching to the hold phase must be selected to a switchover based on stroke or volume.

2. The injection speed set can be obtained from a relative viscosity test; alternatively the known injection speed of the mold operation can be used.

3. Set the holding phase to zero (both the holding pressure and the holding pressure time to zero).

4. Check if the injection of short or incomplete parts could cause problems (e. g. for ejectors, slides, etc.). If there are any problems, calculate the amount of material required to meet the desired cavity-fill percentage. Verify and control that during the metering stage, molten material does not go into the cavity due to back pressure during this stage, which could happen with low-viscosity materials. If this does happen, delay the metering stage until the gates are completely sealed to

avoid erroneous cavity filling inputs. We should get short parts in each cycle and always have a material cushion available.

5. Inject and compare the value of the injection pressure with the value set as the injection pressure limit value. The real mold filling pressure reached can be (a) equal or close to the set injection pressure limit or (b) lower than the set injection pressure limit. In case (a), gradually increase the injection pressure limit in 100 bar increments until the real injection time is constant and no longer decreases. This repeatability of the injection time indicates that the injection pressure is not limiting the process. In case (b) (with the real injection pressure lower than the setting), we are molding without a pressure limit. We must reduce the injection pressure limit setting until the actual filling time increases. At that precise moment, we are intentionally working with a limited injection pressure process. Then, we will increase the specific injection pressure limit in 100 bar increments until the real mold filling time is constant. We will set the injection pressure limit to 150–300 (specific injection pressure) bar above the real injection pressure, or 10–20%.

Table 7.8 Collected Delta P Data*

Cycle	Set Injection Pressure Limit (bar)	Filling Time (s)	Injection Pressure – Real Peak (bar)	Delta P
1	1100	2.4	1050	Pressure limited
2	1200	2.2	1130	Pressure limited
3	1300	1.9	1200	Pressure limited
4	1400	1.85	1280	Pressure limited
5	1500	1.7	1310	Pressure limited
6	1600	1.65	1405	Pressure limited
7	1700	1.6	1447	Time shorter
8	1800	1.6	1450	Filling time repeat
9	1900	1.6	1451	Filling time repeat
10	2000	1.6	1450	Filling time repeat

* In this case, the Delta P value should be 1950 bar (1700 + 250 – the injection pressure limit)

Figure 7.34 shows a sample spreadsheet for entering the test data and obtaining the graph that will tell us when the filling time stops decreasing or potentially starts increasing if we lower the pressure limit and reduce the real injection pressure.

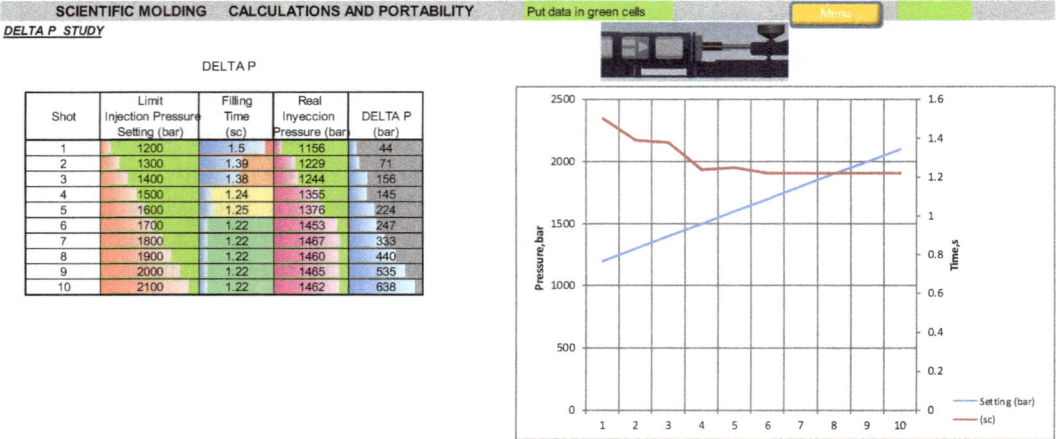

	SCIENTIFIC MOLDING CALCULATIONS AND PORTABILITY		Put data in green cells		Menu

DELTA P STUDY

DELTA P

Shot	Limit Injection Pressure Setting (bar)	Filling Time (sc)	Real Inyeccion Pressure (bar)	DELTA P (bar)
1	1200	1.5	1156	44
2	1300	1.39	1229	71
3	1400	1.38	1244	156
4	1500	1.24	1355	145
5	1600	1.25	1376	224
6	1700	1.22	1453	247
7	1800	1.22	1467	333
8	1900	1.22	1460	440
9	2000	1.22	1465	535
10	2100	1.22	1462	638

Figure 7.34 Example of a Delta P study spreadsheet

Whether your machine is electric or hydraulic, closed-loop or open-loop, your goal must be to keep the filling time constant, which requires additional available pressure. This is the "Delta P" – the difference between the required injection pressure and the machine's available injection pressure.

7.2.8 Gate Sealing Study

The open passage of material through the gate will be available for a certain time which will vary according to the processing conditions (melt temperature, mold temperature, etc.) and the gate dimensions. Once the cavity is filled, the polymer molecules are rearranged. In the case of semi-crystalline materials with a much higher order level, this rearrangement causes a loss of volume occupied by the polymer which it is necessary to compensate for in order that parts with good appearance and mechanical properties may be obtained.

7.2.8.1 Cavity Filling Analogy

As explained in Section 4.8.5, the analogy of cavity filling is that the mold cavity is a room and the polymer molecules that we will put in the cavity during the filling are boxes. During filling, the cavity this is filled with unorganized boxes but, when the molecules are rearranged after, they leave an empty space that needs to be filled with more boxes or molecules, provided that, according to the DuPont analogy, the gate of the room and the runner are open! (See Figure 7.35.)

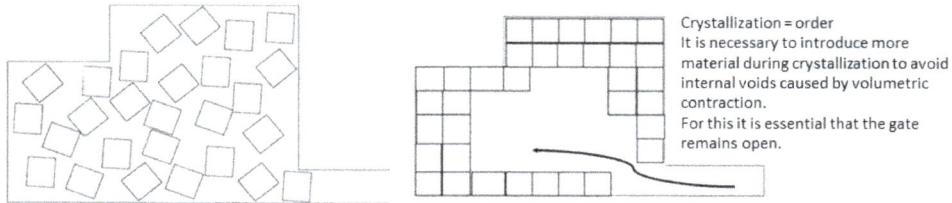

Crystallization = order
It is necessary to introduce more material during crystallization to avoid internal voids caused by volumetric contraction.
For this it is essential that the gate remains open.

Figure 7.35 Left: cavity filling. Right: end of cavity filling; molecule rearrangement; free volume to fill; gate should be open

7.2.8.2 Pack and Hold Stage

It is during this stage that we will define the part dimensions and replicate the surface finish cf the cavity, amongst other things. If this volumetric compensation is not carried out correctly during the packing and holding stage, sink marks or internal voids will appear. There are two possible mistakes at this stage: the holding pressure time is either too short or too long.

- Holding pressure time too short. This does not ensure that the gate is closed, and it can cause backflow of the molten, pressurized material. Imagine we are inflating a balloon. When it is already pressurized, if we stop forcing air in, the air will immediately come out.

  An injection process with a holding pressure lower than the optimum time or an injection process with an unsealed gate means that we have a time-dependent process. Depending on the holding pressure time, we will get parts of different dimensions, properties, weights, etc. In other words, the process will not be consistent and repeatable. By not ensuring that the gates are sealed, the process will be less robust and consistent and the part precision weight, etc. will be negatively affected. This backflow of material is very detrimental to the proper homogeneity of the part's internal structure.

- Holding pressure time too long. This will result in a loss of competitiveness, due to long cycle times and, in the worst-case scenario, additional stress in the areas close to the gates.

  On the contrary, once the gate has been sealed, an excessively long holding pressure time drains energy, due to an excessive cycle time – a waste we cannot afford. I like to see big sink marks in the runners that mold parts correctly. Otherwise, we are manufacturing "precious runners" that no one will appreciate, and no one will pay for.

7.2.8.3 Gate Sealing Time

We need to know the time it takes for the gates to freeze to avoid excessive holding-pressure time and therefore lost productivity. Gate sealing time is determined by:

- Cavity pressure drops (a cavity sensor is required)

- Part weight development control

Cavity pressure drop:

To determine the optimum time for holding pressure through this system, we can set a short holding pressure time so the cavity pressure will drop sharply when the hydraulic pressure ceases. However, if the time is long enough, when we release the injection pressure, the cavity pressure remains and drops slowly, indicating that the gate has been sealed.

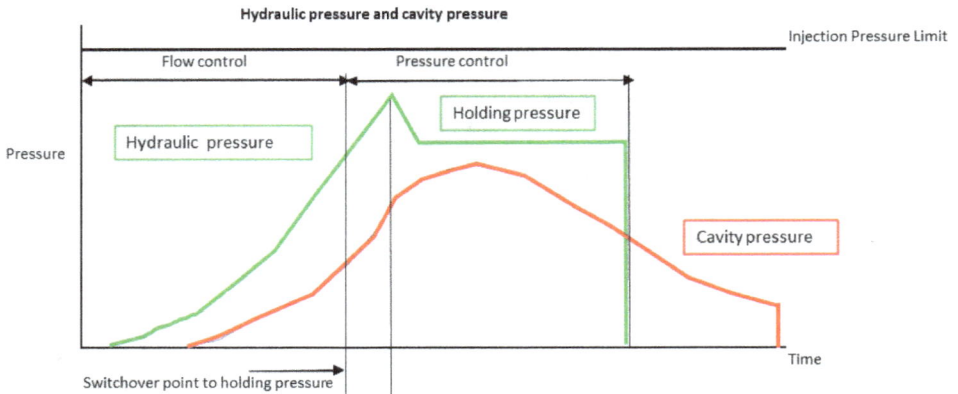

Part-weight control:

To implement this system, it is only necessary to use a bench scale to weigh the molded parts.

The optimum time is when the maximum weight of the molded parts has been reached. Initially, by increasing the holding pressure time, the weight increases, but this eventually stops despite the increase in holding pressure time. At this point, we can conclude that we have frozen the gate and do not continue putting more material into the cavity to compensate for the volume loss during shrinkage. At this point, the weight graph becomes flat.

7.2.8.4 Gate Sealing Study: Injection Machine Configuration

1. Set the injection speed obtained from the relative viscosity study (see Section 7.2.1)

2. Set a cooling time sufficient for obtaining solid parts for ejection from the mold

3. Set holding pressure to the pressure obtained in the process window study (e. g. 50% of the injection pressure needed to fill the cavity)

4. Set the holding pressure time to zero and increase this value while weighing the parts

5. Note that each time you increase the holding pressure time, you need to subtract the same amount of time from the cooling time to ensure that the cycle time remains the same throughout the test

6. Weigh the parts, without taking the runner into account, and check the part weight at each injection time increment

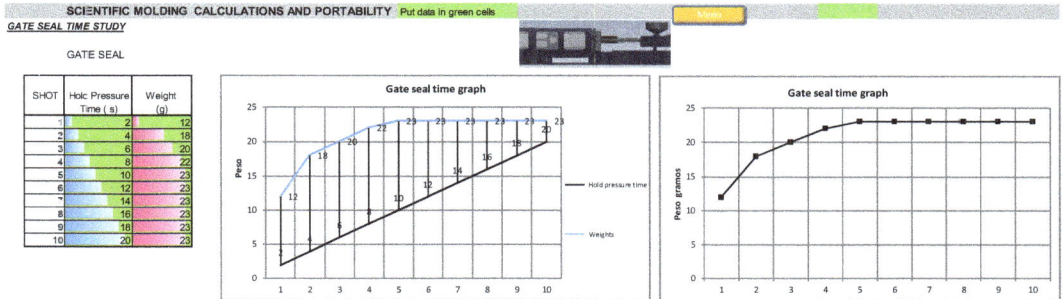

Figure 7.36 Example of a spreadsheet for checking part weight development under different holding pressure times

7. Select the holding pressure time with sealed gates. This value is located in the flat portion of the curve, just after the elbow, where the weights are stable (see Figure 7.37).

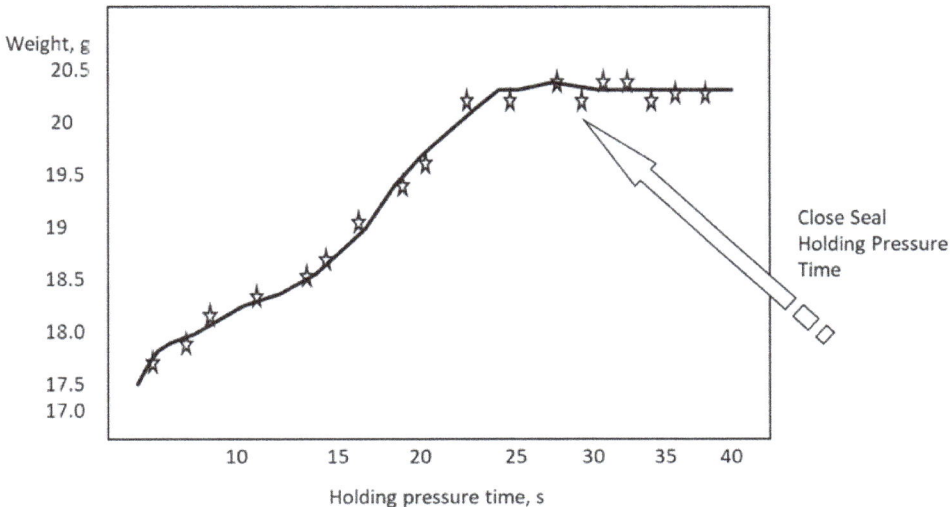

Figure 7.37 Example of part weight development under different holding pressure times

7.2.8.5 Molding with an Open or Sealed Gate?

This decision will depend on the type of part to be molded.

- Sealed gate:
  - Advantages: repeatability and consistency of part dimensions and weights
  - Disadvantages: possible stress in the areas near the gate, especially if the gate is large or if we have hot runners or valve nozzles
- Open gate:
  - Advantages: less stress in the areas near the gate
  - Disadvantages: inconsistencies in part dimensions and weights; high precision of the machine at the switchover point is crucial; time-dependent process

Generally, it is recommended that parts be tested mechanically and in conjunction with downstream processes, such as welding, painting, screen printing, metallizing, and chrome plating, to determine the influence of pressure and holding time.

In summary:

Although the closed or sealed gate affords us greater process repeatability process and avoids possible material backflow, the decision to mold with a closed or an open gate must reflect the part to be molded. Bear in mind that the area closing the gate is always an area with physical properties and chemical resistance different from the rest of the part. It is recommended that the obtained parts be tested with both methods (closed and open gate), to check the mechanical and chemical properties, etc. and to select the type of gate that provides parts with better properties.

7.2.8.6 What if the Part Weight does Not Stop Increasing?

In most cases, we can select the most suitable system: open or closed gate. However, there may be cases where the part weight does not stabilize, for example:

- Large gate area
- Low-viscosity or soft materials (such as TPE, TPV, TPU, PE-HD)
- High holding pressure requirements
- Molding with hot runners
- Molding with sealed nozzle valves in hot runners

Sometimes when carrying out a gate sealing time study, we see that the part weight does not stop increasing. This is a sign that we are pushing some material into the cavity when the polymer starts to cool down (remember that in this stage the polymer does not advance – it is practically stationary – so shear does not contribute to the generation of heat). This situation will leave the area near the gate with a high level of internal stress (displaced material in a semi-solid state) and high molecular orientation. To avoid this situation, we can consider adopting the following methodology:

- Increase the holding time and check the weights that will increase, with increasing tendency. When we mold acceptable parts – either aesthetically, dimensionally or both – we record the part weight as the ideal part weight (shown in Figure 7.38 as position "A").

- Increasing the holding pressure time causes the weight to increase, albeit now with a flatter tendency or proportionality, since we have already pressurized the cavity, and we will reach the maximum part weight (position "B" on the graph in Figure 7.38).

- Reduce the holding pressure little by little until the molded parts' ideal weight ("A") has been reached, which we have recorded as the target weight based on aesthetically and/or dimensionally adequate parts (point "1" on the graph).

- Once the target weight has been reached again by reducing the holding pressure, gradually reduce the holding time of the application (point "2" on the graph).

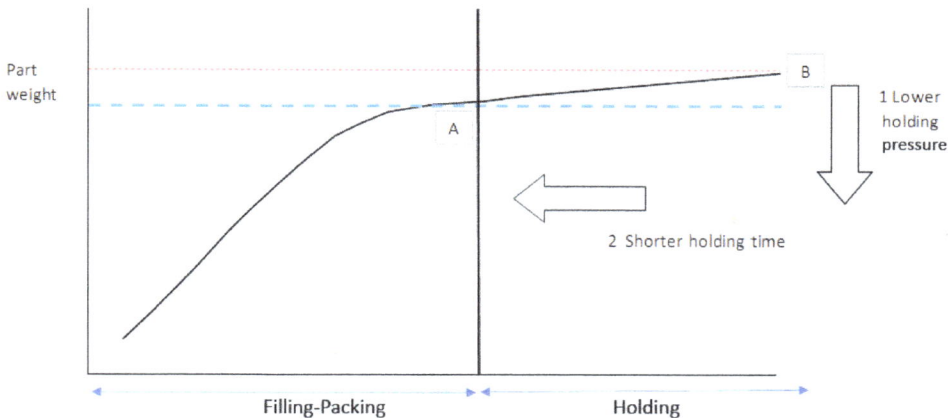

Figure 7.38 Part's weight normal development when hot runners with valve gate system or soft materials are molded during the holding pressure time definition
A: Part weight is acceptable
B: Part weight at the end of holding-pressure time

There will be a moment when the weight starts to decrease. This means that we have reached the bend of the curve but, this time, by reducing the holding-pressure times. At this point, we should add two or three seconds to the set holding-pressure time to ensure that we are in the "safe" portion of the curve. At this point, with the target weight ("A") and the holding-pressure time selected, we seal the gate without increasing the weight. Therefore, no residual stress will be created by the injection of semi-solid material into the cavity.

7.2.8.7 What about Hot Runners or Valve Gate Nozzles?

In this case, carrying out a process window study can help to select the right holding-pressure time in molds with hot runners or valve gate nozzles. This process window can be dimensional, aesthetic, or both, depending on the type of part, sector, etc. A typical procedure is as follows:

1. Set the holding pressure to a low value, for example 250 bar. With this holding pressure make a sequence of shots with increasing holding pressure times (e.g. from 4 to 10 seconds).

    Check the quality of the parts molded (i.e. are ther any sink marks, flasjes, etc. or are the parts adequate?).

2. Then progressively increase the holding pressure at regular intervals, for example in Figure 7.39 from 200 to 650 bar, performing the tests with different holding pressure times, for example from 2 to 7 seconds in Figure 7.39.

Repeating these steps increases the holding pressure to levels that produce defective parts over the range of holding pressure times. With this information, it is possible to draw up a mold area diagram or process window (see Figure 7.39 and the explanation in Section 7.2.9). Selecting a combination of holding pressure and holding pressure time in the central area of the generated process windows yields the most robust and consistent process. As for the valve gate, the selected holding pressure time would not be the time that gives us weight stability, but a time that provides us with aesthetic and dimensional quality, regardless of the weight.

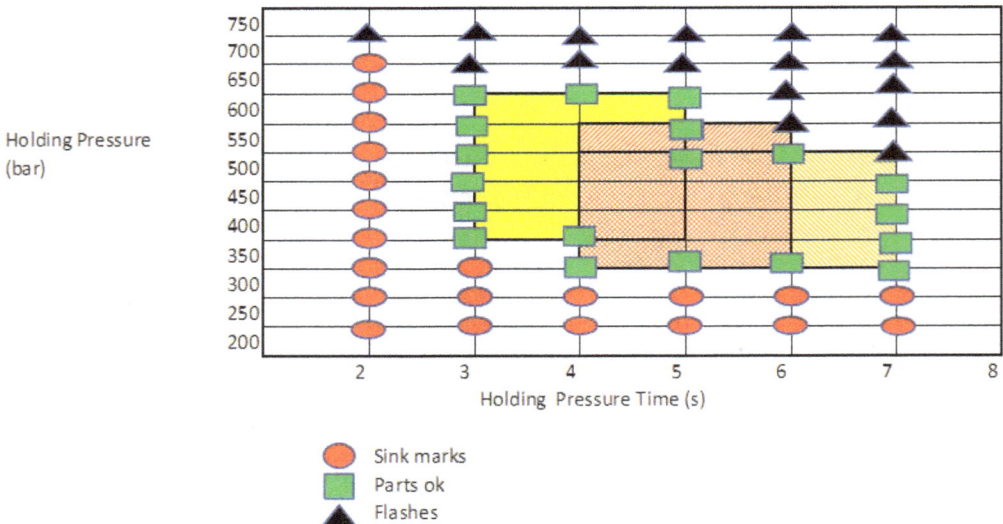

Figure 7.39 An example of a dimensional and/or aesthetic process window

7.2.9 Process Window Determination (for the Holding Pressure Phase)

Filling of the cavity with molten plastic material proceeds out in three distinct stages. The first is volumetric or dynamic cavity filling. In this stage, we control the cavity filling speed and the mold and the melt temperatures. The second stage is the packing pressure stage or volumetric shrinkage compensation that occurs as the material cools down within the mold. In this stage, we control the holding pressure and its application time. This second stage is called the packing stage and it determines the weight and dimensions of the molded parts. If this holding pressure is inadequate, we will obtain incomplete parts, with sink marks, reduced dimensions and weight, voids, etc. Conversely, if the holding pressure is excessive, we could obtain oversized parts, with flash, internal stress, complicated ejection, etc. Finally, the third stage is the holding stage for ensuring that the gate is closed or sealed by cooling under the holding pressure.

To optimize the packing and holding pressure stages, we can apply the "process window" methodology or "diagram molding", where the "area" in which our process (the machine, mold, material and parameters) will produce acceptable parts is determined (Figure 7.40). The goal is to have the "area" of the process window as large as possible so that we will have a most robust and consistent process.

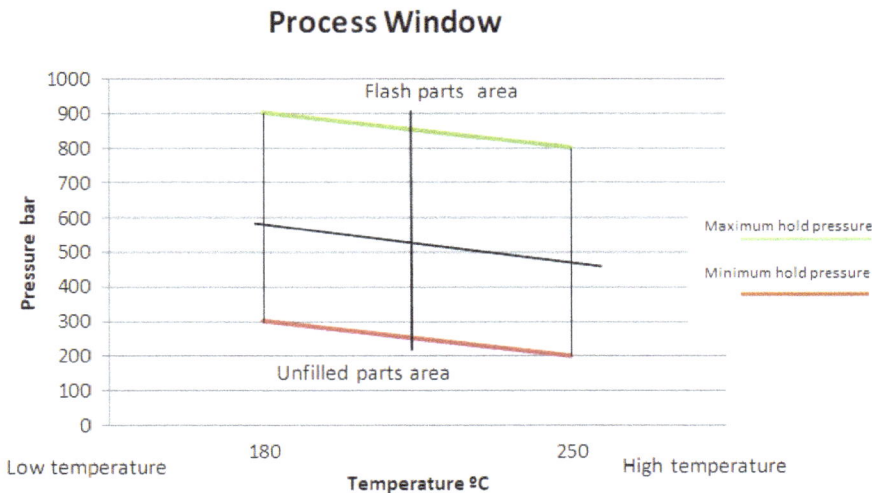

Figure 7.40 Example of a process window study

In Figure 7.40, we can see that the parts produced outside the limits of the process window have flash or short parts to an extent depending on the pressure of the holding pressure stage. If exceeding the limits affects the melt temperature, the parts may have flash, due to degradation, loss of mechanical properties, or unmelted and inhomogeneous melt, and so the parts will be unacceptable. If the defined process is as close to the center of the process window as possible, it will be much more robust and consistent and it will absorb small variations that may occur during manufacturing.

The analogy here would be that our injection unit is like a sharpshooter, and each shot or cycle represents one gunshot. Our process window would be the target at which our process shoots. Each injection cycle that produces defective parts represents shots off-target. The larger the process window (or target), the more shots we get on target, resulting in more acceptable parts and better productivity ratios.

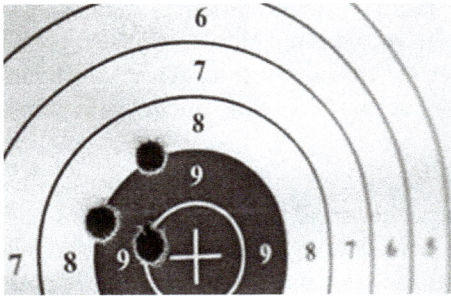

7.2.9.1 Process Window Study: Injection Machine Configuration

1. Set the melt processing temperatures to the polymer manufacturer's lower recommendation

2. Set the injection speed obtained from the in-mold viscosity test, linearity test, and filling balance test, or the injection speed used to fill the mold

3. Set the holding pressure time and holding pressure to zero

4. Set a cooling time slightly longer than necessary

7.2.9.2 Process Window Study: Steps

1. Start and adjust the switchover point to the holding pressure stage at 95–98% of the total cavity filling

2. Run several consecutive cycles to stabilize the process

3. Set the holding pressure time longer than necessary to ensure that the gates are frozen

4. Parts molded will have sink marks and low aesthetic and mechanical properties

5. Increase holding pressure in pressure intervals until acceptable parts are obtained

6. This pressure and temperature are the points shown on the graph in Figure 7.40:
   - Low temperature
   - Low pressure at the bottom left corner

7. Increase the holding pressure intervals and record when the pressure is excessive (parts with flash, deformations, mold grip, etc.). These should be noted:
   - Low temperature
   - High pressure (top left corner) (see the graph in Figure 7.40)

8. Set the melt processing temperatures to the polymer manufacture's higher recommendation and increase the holding pressure from zero in pressure intervals until acceptable parts are obtained

9. This time, the following points found are:
   - High temperature, low pressure
   - High temperature, high pressure
     (in the top and bottom right corners)

   By joining the four sub-points given above (the four combinations of low/high temperature and low/high pressure), we get the process window or diagram of molding area.

10. Set the process parameters for melt temperature and holding pressure as close to the center of the defined area as possible

The process window indicates the amount of variability that the defined process can absorb before it starts injecting or producing defective parts (i. e. working outside the process window). If the process window is narrow (and therefore not very robust), it is advisable to consider how the area might be increased. These actions are determined by defects or problems encountered during this process. With this test, we can therefore not only determine the window process area, but we can also easily identify which actions should be taken to increase this area.

7.3 Further Scientific Molding Tools or Studies for Checking and Improving the Injection Molding Process

The following studies or scientific injection molding tools can be used to subsequently check the robustness and consistency of the defined injection process and to optimize the results obtained previously.

7.3.1 Shear Rate at the Gate Study

Typically, the gate is the most restrictive passageway for material in transit from the machine to the cavity. When a mold is machined, the gates are very often small because it is easier and cheaper to increase the gate size than to reduce it (welding, adjustments, etc.), if necessary. Unfortunately, molders sometimes decide to adjust the injection parameters to a gate size that is inadequate instead of modifying the gate size. This leads to molded parts with less-than-optimal conditions or parameters (e. g. a higher-than-recommended melt temperature or injection speed, a smaller process window or molding area, or even worse, molded parts with a higher shear rate than allowed by the material being injected).

All polymers have maximum shear rate values that should not be exceeded if the risk of material degradation is to be avoided. Exceeding these recommended values will result in molecular degradation of the polymer. If we exceed the values, we can cause an irreversible loss of material properties, in addition to aesthetic defects, such as cracks, marks in the gate, gases, voids, brightness, etc.

The following equations can be used to calculate the correct gate dimensions, taking into account the maximum shear allowed for the material used. The equations give us the shear rate or friction generated by passing a molten polymer through a circular or rectangular gate.

Circular gate:

$$\text{Shear rate} = 4Q/3.142 \times r^3$$

Rectangular gate:

$$\text{Shear rate} = 6 \times Q(\text{gate length} \times \text{gate width}^2)$$

where Q = flow (cm^3 per second) and r = channel radii

When we use these equations, the results must not exceed the maximum shear rate values recommended for the material being molded (Table 7.9).

Table 7.9 Maximum Shear Rate Allowed for Some Materials

Material	Max. Shear Allowed (s^{-1})
PP	100,000
PC	40,000
ABS	50,000
ABS chrome plating grades	30,000
PMMA	40,000

Material	Max. Shear Allowed (s^{-1})
PES	50,000
PET	50,000
PBT	50,000
PA 12, PA 612	60,000
PA 66	60,000
PA 6	60,000
PE-LD	40,000
HIPS	40,000
GPS	40,000
EVA	30,000
POM	40,000
PPO	35,000
PPS	50,000
PSU	50,000
PUR	40,000
PVC	20,000
SAN	40,000

The equation for circular gates includes:

- Q = flow rate; this can be calculated from the cavity volume and filling time values (cm^3/s)

- r^3 = radius of the gate diameter (mm) raised to the power of 3

We can deduce that a small change in the diameter heavily influences the result of the shear applied.

The equation for rectangular gates includes:

- Q = flow rate (cavity volume and filling time)

The geometric values of the rectangular gate are also included (i. e. width and depth or thickness (mm)). The thickness is squared, so the depth of the gate heavily influences the final shear calculation.

When we use these equations with different combinations of filling time and gate area, the maximum friction and shear values recommended by the polymer manufacturer's data should not be reached. We can calculate the shear and modify the filling time (flow rate), gate area, etc. to check the shear development using the different combinations analyzed.

In the example of the calculation grid in Figure 7.41, by defining the range of diameters and injection times, we can see multiple results for these combinations and clearly determine the results that achieve data below the maximum recommended shear of the material (green color) and the combinations that reach shear above the recommended one (red color). With these results, we can select which filling time (injection speed) and gate diameter we will use in our process to ensure the correct shear rate.

Table 7.10　Data for Calculation (Example of Circular Gate Data)

Data Description for Shear Rate on Circular Gate Calculation	Data	Units
Maximum shear rate	60000	s^{-1}
Part weight	6.32	g
Material density at 23 °C	0.9	g/cm^3
Melt density	0.765	g/cm^3
Volume	8.2614379	cc

Filling time (sc) →		0.6	0.7	0.8	0.9	1	1.1
CIRCULAR GATES	Gate → diameter mm	1	140268.06	120229.8	105201	93512.04	84160.834	76509.85
		1.1	105385.47	90330.4	79039.1	70256.98	63231.281	57482.98
		1.2	81173.644	69577.41	60880.23	54115.76	48704.187	44276.53
		1.3	63845.27	54724.52	47883.95	42563.51	38307.162	34824.69
		1.4	51118.097	43815.51	38338.57	34078.73	30670.858	27882.6
		1.5	41560.906	35623.63	31170.68	27707.27	24936.544	22669.59
	maximum SHEAR s⁻¹		60000	60000	60000	60000	60000	60000

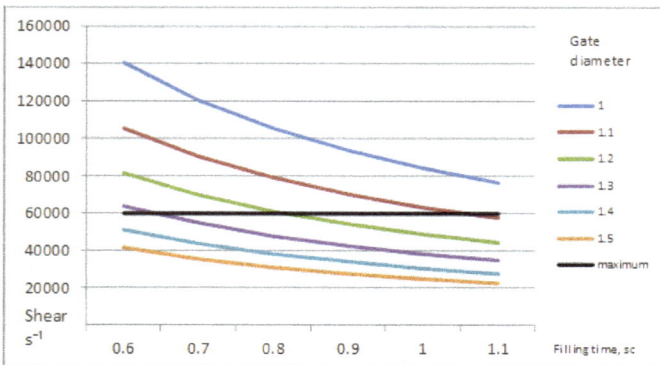

Figure 7.41　An example of circular gate graph and shear calculation (see data in Table 7.10)

In the graph in Figure 7.41, we can see that the values calculated with combinations of filling time and gate diameters yield shear results below the black line representing the maximum shear for the material in question (e. g. ABS with a maximum shear of $50,000\,s^{-1}$). The black line in the graph is the maximum allowable shear for the material under analysis (e. g. $50,000\,s^{-1}$) and the curves for different gate diameters (from 0.8 mm to 1.3 mm) combined with different filling times (from 1 to 2 seconds).

In the example shown in Figure 7.41, we can conclude that filling times of less than 1.5 seconds with a gate diameter of less than 1.1 mm (purple color) exceed the maximum allowable shear. For the rest of the diameters (1.2 mm and 1.3 mm), the whole range of filling times analyzed (from 1 to 2 seconds) is acceptable because it does not exceed $50,000\,s^{-1}$ (maximum shear).

This analysis is very useful when we want to check that our process and combinations of filling time and gate dimensions do not apply an excessive shear rate to the material. To check this, we only need to know the volume of our cavity (cm^3), the filling time (s) and the gate area (mm^2). This data will allow us to evaluate possible modifications to gate dimensions or injection speed, if necessary, to reduce the shear applied; see Figure 7.43.

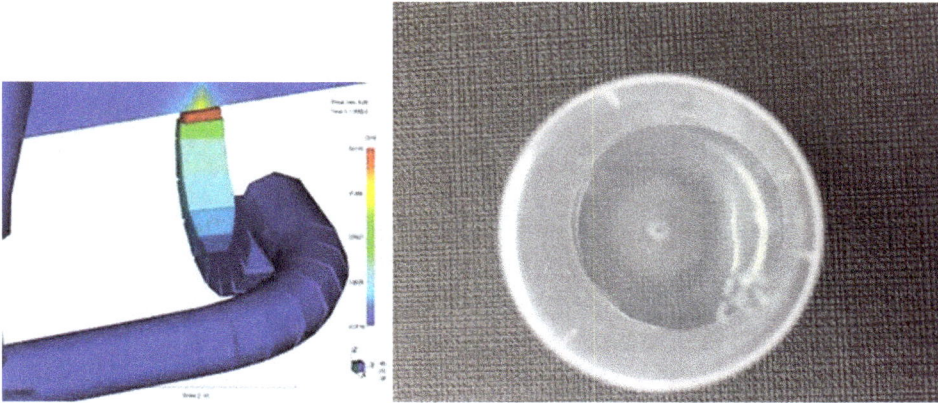

Figure 7.42 An example of the gate restriction effect

SCIENTIFIC MOLDING CALCULATIONS AND PORTABILITY Put data in green cells

SHEAR STRESS IN CIRCULAR GATES STUDY

CALCULATION DATA CONFIGURATION		
Minimum filling time	0.6	sc
Step difference time between calculations	0.1	sc
Minimum gate diameter	1	mm
Step difference diameter between calculations	0.1	mm

MATERIAL	PA6

DATA FOR CALCULATION		
Maximum shear rate	60000	s^{-1}
Part weight g	6.32	g
Density mat 23°C	0.9	g/cm^3
Melt density	0.765	g/cm^3
Volume	8.2614379	cm^3

	Maximum shear recommended	s^{-1}
PP	100,000	
PC	40,000	
ABS	50,000	
ABS plating grades	30,000	
PMMA	40,000	
PES	50,000	
PET	50,000	
PBT	50,000	
PA 12 PA 612	80,000	
PA 66	60,000	
PA6	60,000	
LDPE	40,000	
HIPS	40,000	
GPS	40,000	
EVA	30,000	
POM	40,000	
PPO	35,000	
PPS	50,000	
PSU	50,000	
TPU	40,000	
PVC	20,000	
SAN	40,000	

CIRCULAR GATES

Gate diameter mm

Filling time (sc)		0.6	0.7	0.8	0.9	1	1.1
	1	140268.06	120229.8	105201	93512.04	84160.834	76509.85
	1.1	105385.47	90330.4	79039.1	70256.98	63231.261	57482.96
	1.2	81173.644	69577.41	60880.23	54115.76	48704.187	44276.53
	1.3	63845.27	54724.52	47883.95	42563.51	38307.162	34824.69
	1.4	51118.097	43815.51	38338.57	34078.73	30670.868	27882.8
	1.5	41560.908	35623.63	31170.68	27707.27	24935.544	22669.59

maximum SHEAR s^{-1}

	60000	60000	60000	80000	60000	60000

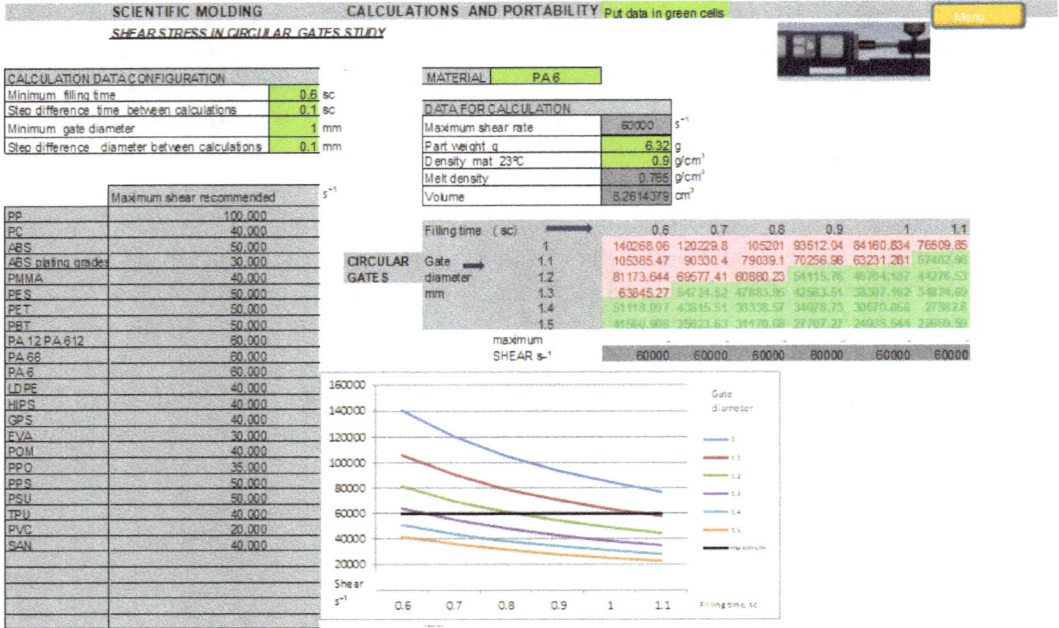

SCIENTIFIC MOLDING CALCULATIONS AND PORTABILITY Put data in green cells

SHEAR STRESS IN RECTANGULAR GATES STUDY

MATERIAL	PPO

CALCULATION DATA CONFIGURATION		
Minimum filling time for calculs	0.6	sc
Step difference time between calculations	0.1	sc
Ratio width/depth	2	times
Gate depth	1	mm
Deep Step difference between calculations	0.1	mm

DATA FOR CALCULATION		
Maximum shear rate	35000	s^{-1}
Part weight	6.32	g
Density material	0.9	g/cm^3
Melt density	0.765	g/cm^3
volume cm	8.261438	cm^3

	Maximum shear recommended
PP	100,000
PC	40,000
ABS	50,000
ABS plating gr	30,000
PMMA	40,000
PES	50,000
PET	50,000
PBT	50,000
PA 12 PA 612	60,000
PA 66	60,000
PA6	60,000
LDPE	40,000
HIPS	40,000
GPS	40,000
EVA	30,000
POM	40,000
PPO	35,000
PPS	50,000
PSU	50,000
PUR	40,000
PVC	20,000
SAN	40,000

RECTANGULAR GATES

mm Gate dimension width/depth		Gate area mm	Filling time s	0.6	0.7	0.8	0.9	1	1.1
	1	2	2	41307.19	35406.162	30980.392	27538.13	24784.31	22531.19
	1.1	2.2	2.42	31034.7	2660.1.174	23278.02717	20689.8	18520.82	16928.02
	1.2	2.4	2.88	23904.62	20489.677	17928.46769	15936.42	14342.77	13038.89
	1.3	2.6	3.38	18901.63	16116.586	14101.22538	12534.42	11280.98	10255.44
	1.4	2.8	3.92	15053.64	12903.12	11289.23038	10035.76	9032.184	8211.077
	1.5	3	4.5	12239.17	10490.715	9179.375454	8169.446	7343.5	6675.909

Depth, mm	Width, mm	Max. Shear s^{-1}

Maximum Shear s^{-1}

35000	35000	35000	35000	35000	35000

Q cm^3/sg

13.76905	11.80254	10.32479739	9.17937	8.261438	7.510398

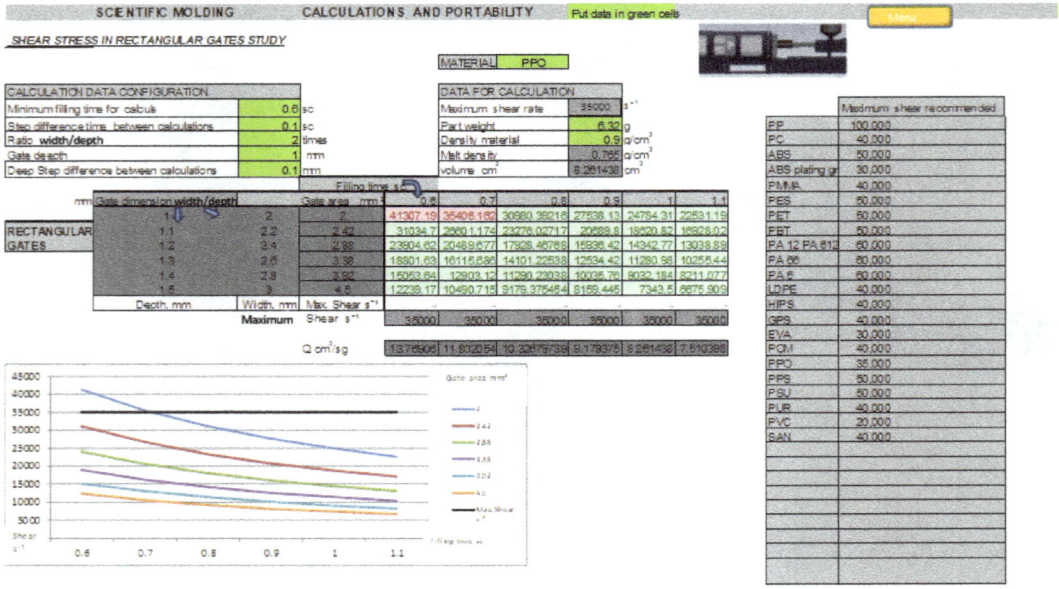

Figure 7.43 An example of spreadsheets that relate the equations to the graphs for circular and rectangular gates

7.3.2 Cooling System Study

The objective of the cooling system (from the impeller, pump or temperature control unit, pipes, hoses, flow meters, amongst others, to the channels designed in the mold for the coolant to flow), is to remove heat from the molten material once the cavity has been filled and pressurized, so that the material undergoes the volumetric shrinkage that allows us to demold the parts and is rigid and strong enough to withstand the force exerted by its ejection from the mold.

In Figure 7.44, we can see the mold temperature cycles in each shot or cycle. As the molten material fills the cavity, the latter receives a high thermal pulse that quickly raises the temperature of the steel. The cooling system then removes heat from the mold through the coolant flow, cooling the molten polymer inside the mold. The rate of cooling depends on the temperature of the coolant and the flow rate through the mold. In this system, the mold works as a heat exchanger. The problem is that this heat exchanger is a large mass of steel with a (very) slow thermal response. Like any heat exchanger, its performance is the ratio between the temperature of the mold and the flow conditions of the circulating coolant (liters per minute).

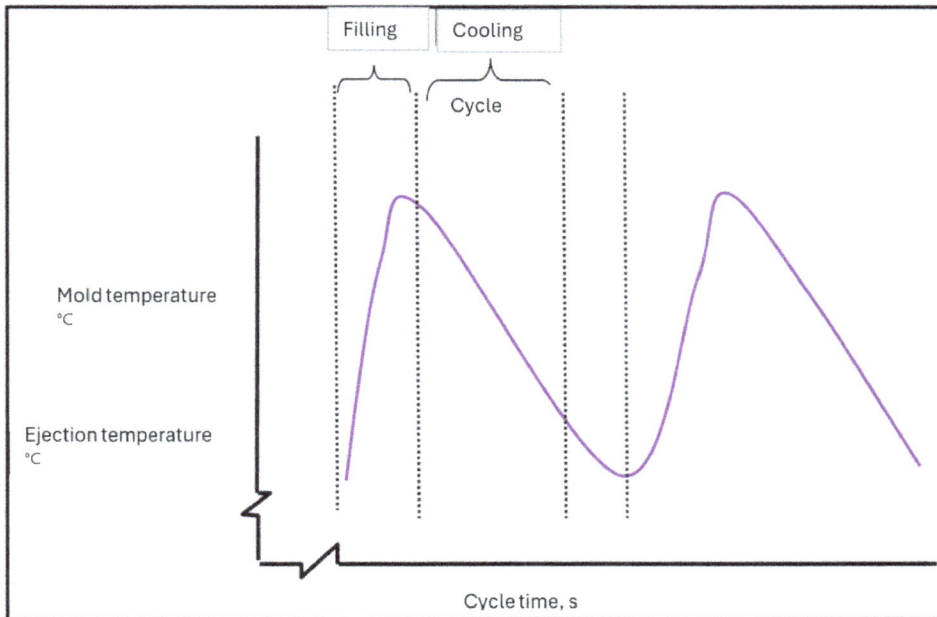

Figure 7.44 Mold wall temperature during an injection cycle (Source: Eurecat)

Coolant can flow in three different states: laminar, transitional, or turbulent. Laminar flow means that the coolant molecules (often water) move in an aligned manner so that their ability to exchange heat with the circuit is low. Only the molecules closest to the cavity wall can remove and exchange heat, while those closest to the core will engage in little heat exchange. In contrast, the turbulent flow involves a large internal turbulence in the layers and molecules of the coolant, so that all the coolant molecules in contact with the walls of the cooling system act as coolant. This makes the effectiveness of the turbulent flow up to four times greater than laminar flow.

The Reynolds number (Re) is used to calculate the type of flow of a circulating liquid. It is a dimensionless unit that indicates the type of flow of a liquid circulating in a piping system. It is defined as:

- Laminar flow: Reynolds number < 2500

- Transitional flow: Reynolds number between 2500 and 4000

- Turbulent flow: Reynolds number > 4000

For laminar flow (see Figure 7.45), the heat transfer coefficient is approximately 500 kcal/h m^2 °C, while if we move to turbulent flow, this can increase up to more than 2000 kcal/h m^2 °C – more than four times the heat transfer capacity. This can have a big impact on cycle time and therefore productivity and costs.

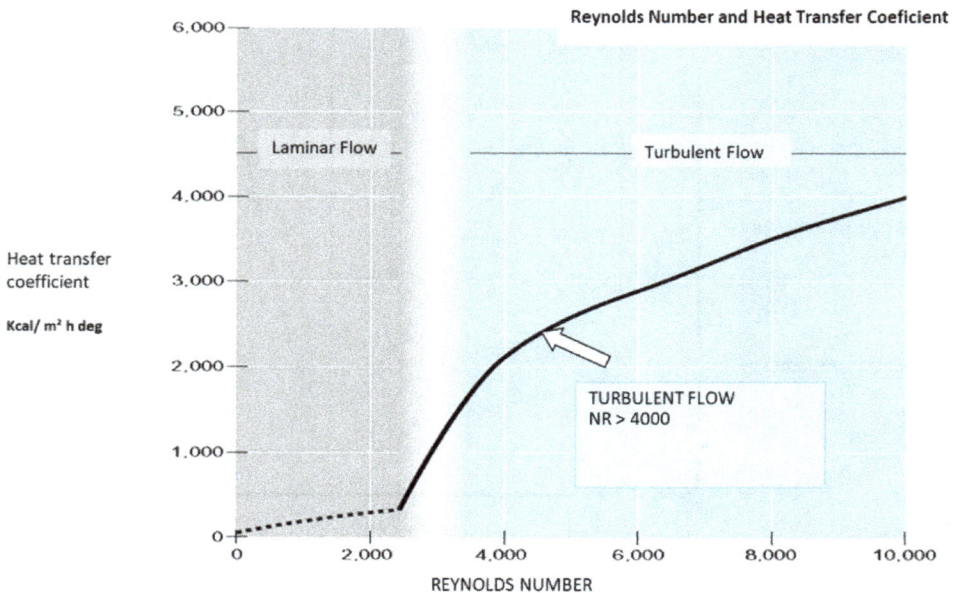

Figure 7.45 Reynolds number and heat transfer coefficient and coolant flow graph

The calculation of the Reynolds number (Re) involves the kinematic viscosity of the water at different temperatures, the coolant flow rate, and of course the dimensions of the cooling system:

$$Re = VD\rho/\mu$$

where V is the speed, D the diameter, ρ the fluid density, and μ the kinematic viscosity.

Cooling with turbulent flow improves cavity temperature stability. Studies have been carried out (Burger & Brown, *https://www.smartflow-usa.com/measuring-turbulent-flow/*) consisting in placing temperature sensors in the mold and simulating the heat input that occurs in each cycle as the cavity is filled with molten plastic. It was observed that, as the coolant flow rate increases, the cavity temperature decreases, but there is a wide range of flow rates that affect the cavity temperature. It is therefore a process in which the cavity temperature depends on the coolant flow rate.

However, as the Reynolds number increases and the flow becomes turbulent, the cavity temperature is stable, regardless of the coolant flow rate. As can be seen in Figure 7.46, in turbulent flow, the cavity temperature is more consistent or flat. In these cooling conditions, the flow rate has virtually no effect on the mold temperature, resulting in a more robust and repeatable process.

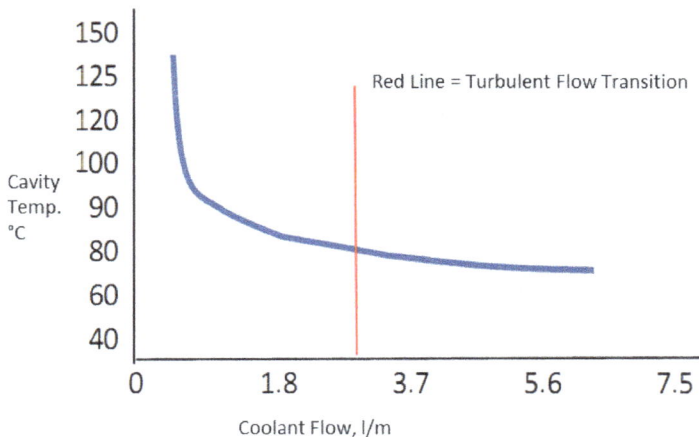

Figure 7.46 Cavity temperature stabilization with turbulent flow

At flow rates below turbulent flow, the temperature of the steel is less stable and highly dependent on the flow rate. During turbulent flow, however, the cavity temperature becomes more stable and less dependent on the flow rate through the mold.

In short, we can calculate the coolant circulation flow with this tool to optimize it. In turbulent flow, we will obtain a better heat transfer coefficient for the cooling system, which will result in better cycles and more stability in the cavity temperature.

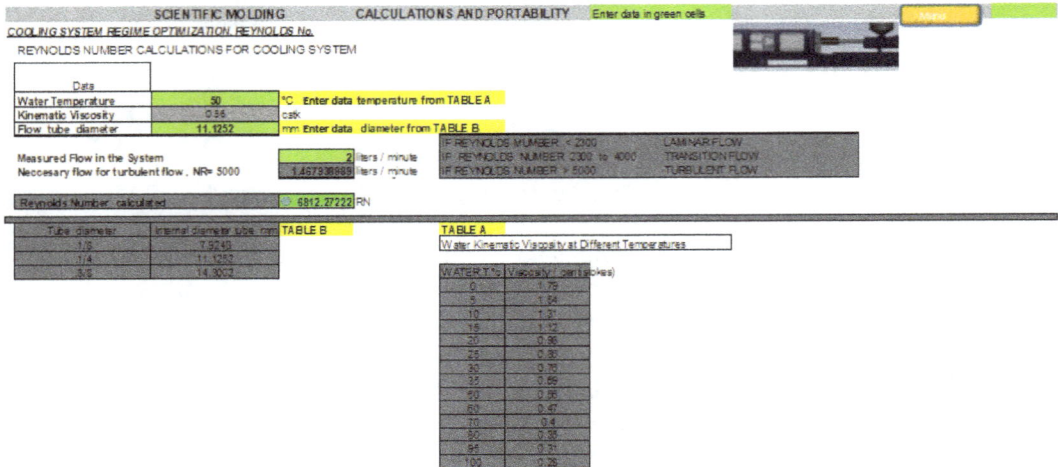

Figure 7.47 An example of a spreadsheet for a mold cooling system study

7.3.3 Cooling Time Study

This cooling time study takes dimensions into account. By definition, the cooling time is the time required for a part wall to reach a temperature that allows it to be demolded undeformed or with dimensional accuracy. To determine the ideal cooling time, we can take two or three critical dimensions of the part and analyze the influence of the cooling time on these final dimensions. By plotting the results of these time–dimension tests, we can determine if there is a direct correlation between dimensions and cooling time. If there is, we can determine the cooling time or cooling time range that will produce acceptable parts (Figure 7.48).

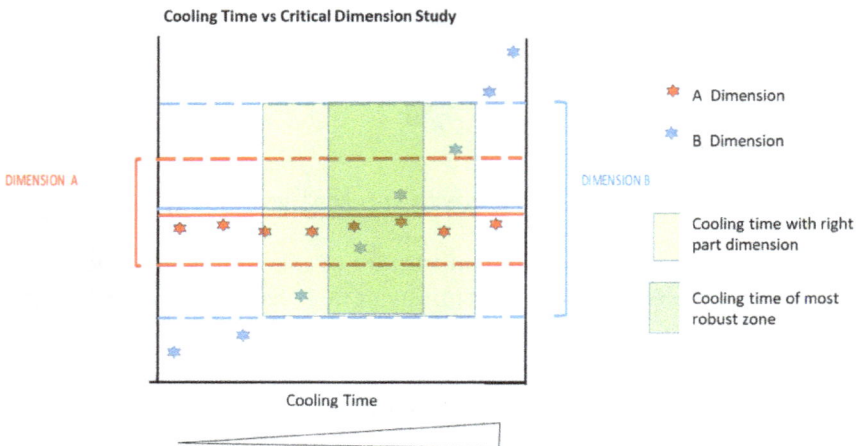

Figure 7.48 An example of a cooling time study. It can be seen that the cooling time has no influence on DIMENSION A, but a great influence on DIMENSION B

A short cooling time will result in greater shrinkage outside the mold at constant pressure (atmospheric). A long cooling time will lead to less shrinkage, which will occur inside the mold under decreasing pressure. It is also important to note that shrinkage inside the mold is restricted by the mold nozzles, mold walls, slides, etc., but outside the mold, the part shrinkage is unrestricted, due to a lack of mechanical retention.

Cooling System Study Procedure

Once the injection parameters have been defined:

1. Perform three injection shots, each with a different cooling time (from short to long)

2. Check the critical part dimensions with the different cooling times

3. Plot the dimensions against the cooling time

4. Determine whether there is a direct correlation between the dimensions and cooling time and which time(s) allow us to mold parts within part dimensional tolerances

5. Select the cooling time for final testing

6. Perform 30 shots to ensure that the molded parts dimensions are statistically robust and capable, Cpk, etc.

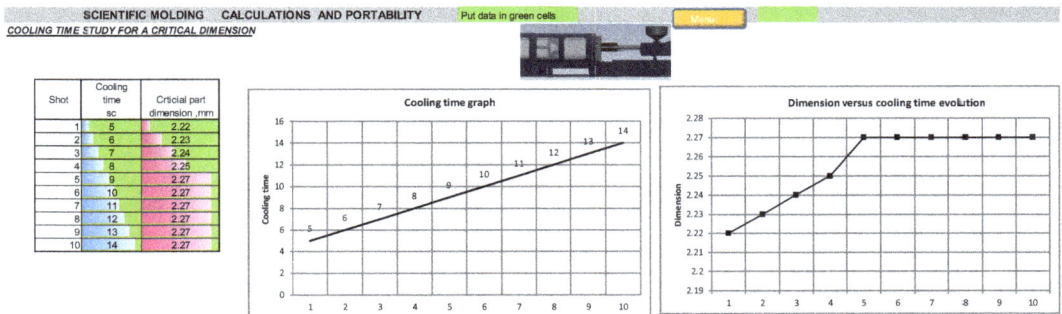

Figure 7.49 An example of a spreadsheet for analyzing the cooling time for a critical dimension

7.3.4 Study of Material Residence Time in the Injection Unit

Plastics must be able to withstand processing temperatures without degradation. This heat resistance is enhanced by certain additives and heat stabilizers that help to improve this resistance. However, the resistance is subject to time constraints.

The plastic is heated during the transformation period, that is, from the moment the pellets enter the throat of the injection unit until they become a molded part. Heat energy applied for too long leads to polymer degradation. This means an irreversible loss of molecular weight and the consequent loss of properties of the manufactured parts. This thermal degradation is closely related to temperature and time – the higher the process temperature at which the plastic is being molded, the shorter the maximum time that the plastic can remain at that temperature.

Figure 7.50 shows the degradation curves for a PBT. For example, at 290 °C we have a maximum residence time of 8 minutes, whereas at 300 °C we would have barely 2 minutes for the same material. If the material is flame resistant, the thermal degradation based on residence time is faster. In this case, at 280 °C, the maximum residence time for the flame-resisant material would be 4 minutes, and only 2 minutes at 290 °C.

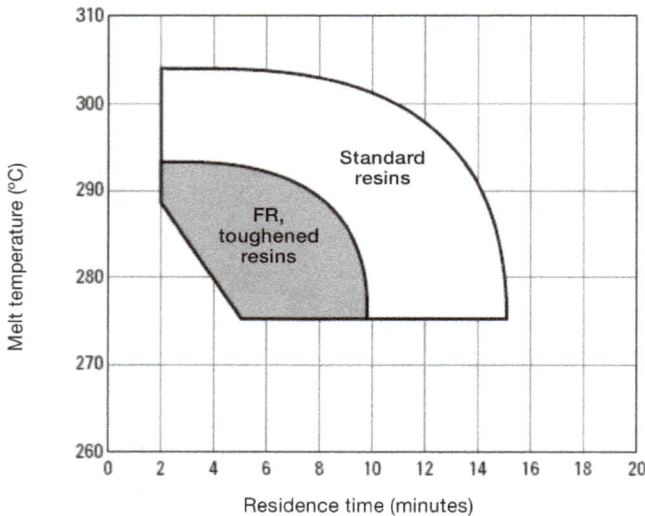

Figure 7.50 An example of a melt temperature vs residence time graph for a PBT (Source: DuPont)

We must therefore know the residence time of the material in the injection unit, which will help us to know if degradation will occur or, on the contrary, if the residence time is correct and there will be no thermal degradation of the material. The equation used to calculate the residence time is as follows:

Residence time (min) =
(maximum metering weight/injection shot weight) × (cycle time/60)

Looking at the factors involved in this equation, it is easy to know the total weight of the shot and the cycle time, but it is not easy to calculate how much material weight is in the injection unit at the maximum metering stroke. Do not confuse this factor with the maximum weight that we can inject with the machine. This is not the weight of the material inside the barrel. In fact, along the entire length of the screw, there is also material in the threads, and the free volume in the threads decreases according to the screw compression ratio. Furthermore, in the screw the material has different densities, solids, semi-solids, melts, etc. Calculating the weight of the material that can fit inside the screw is highly complex.

For this reason, the calculation has been simplified to the following equation:

Residence time (min) =
((screw diameter(mm) × 8) × cycle time(s))/net metering stroke(mm)

The spreadsheet in Figure 7.51 for calculating residence time also provides calculated data on the injection unit utilization and the metering stroke / screw diameter ratio.

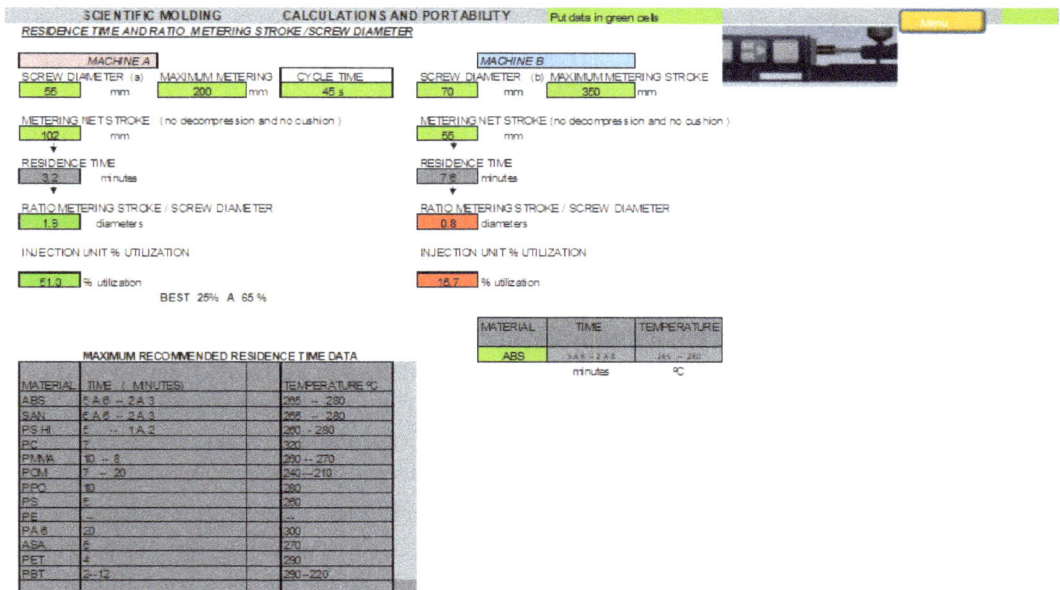

Figure 7.51 An example of a spreadsheet for calculating the residence time

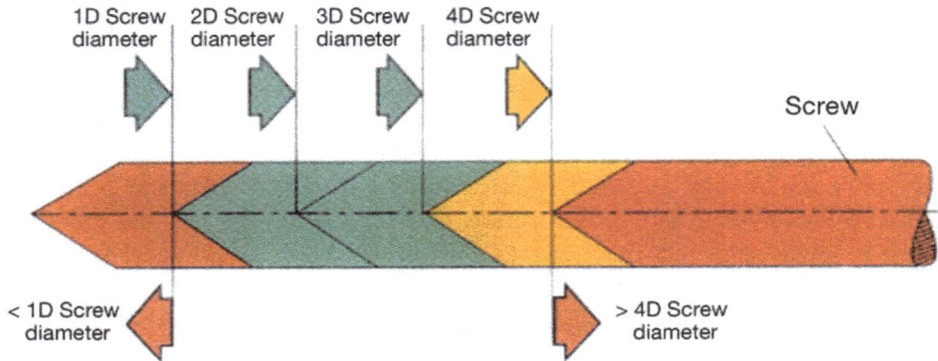

Figure 7.52 Recommendations for the metering stroke / screw diameter ratio

In summary, the residence time of the material must be checked to see if it is shorter than the maximum residence time recommended by the polymer manufacturer or if, on the contrary, it is longer. In the latter case, the process will cause some thermal degradation of the material and loss of properties.

It is also advisable to carry out this study and calculation when there is suspected degradation, breakage of parts, fragility, streaks, etc., or at any time or situation when we want to ensure that we are not degrading the plastic by prolonging the residence time. As a rule of thumb, the longer the residence time, the lower the injection unit temperature required to avoid thermal degradation.

7.3.5 Filling Time Repeatability Study

The filling time repeatability study is a method to check the stability and repeatability of the process at the defined filling time. As introduced in Section 3.1, the mold filling time – the entire injection process's most critical time – is the consequence of the melt flow speed during the filling of the cavities, so its repeatability is very important. Variations in filling time or injection time indicate variations in injection molding parts.

It is critical because during this time the material fills the cavities, and during the filling phase we define many of the molded parts' properties, such as structure, crystallinity, viscosity, entrapped air, gases, peak pressure in the cavity during filling, part temperature range in the cavity, and the range of pressures in the cavity

The injection machine must meet the set injection speeds and therefore the expected injection time. To do this (see Section 7.2.2 and Section 7.2.9), it is essential to have defined a correct Delta P and not to inject with a pressure-limited process. With a correctly defined Delta P, the molder must repeat the filling time regardless of the viscosity of the injected material. As for this time repeatability, there are authors who limit its range of variability to ±0.04 seconds. In my opinion, an acceptable maximum variability would be in the range of 2–3%, all depending on the sector and the required part accuracy.

By frequently carrying out this study, we can check the machine's level of repeatability, so that we can check calibration and machine maintenance if there is any variation at a given moment. To carry out the filling time study:

1. Set the injection pressure limit above the real required injection pressure data to ensure that the process is not pressure limited.

2. Perform 10 injection shots in a row at the selected mold filling speed in line with the studies presented in this chapter.

3. Collect the mold filling times and use these values to calculate the range, the slowest and fastest injection times, and the percentage of time variability. We can use an Excel spreadsheet to do this, for instance (Figure 7.53).

Figure 7.53 Example of a spreadsheet for studying the precision and repeatability of the injection molding machine from a filling time standpoint

This study checks the injection molding machine's ability to repeat the injection time and the potential speed variability during the production of part series. The smaller the time variability, the greater the precision of our machine. In addition, the values obtained will serve as a reference for subsequent tests to assess whether there are increases in the variability so that, if necessary, we can recalibrate the machine to return to the usual values.

7.3.6 Screw Tip or Check Ring Valve Sealing Study

The screw tip or check ring valve is one of the most critical elements of the injection unit operating system. As explained in Section 3.4, the screw tip is the mechanical element that prevents the molten plastic from returning to the rear of the injection unit during the injection phase and allows the material to advance towards the front of the barrel during the metering phase. It is also a non-return valve that must seal the molten plastic to ensure that the injection process is repeatable. This element must be hermetically sealed when the injection pressure is applied and the screw moves forward; otherwise material will leak, which will cause significant differences between shots, a lack of repeatability and a robust process, as well as increased scrap.

In the earliest injection molding machines, the screw was not used to melt and homogenize the molten material, but rather a "torpedo" system applied shear to the material as it passed through the injection barrel, in addition to electrical resistances. The first reciprocating screw was patented and used in injection molding machines in 1956.

With the implementation of the screw, it became necessary to design and apply the non-return valve or screw tip. Today's injection molding machines are equipped with advanced technology, such as non-contact position detectors, ultrasonic sensors, electronic displays for control and monitoring, state-of-the-art servomotors and servo valves, ceramic heaters, etc. However, plasticizing and sealing still rely on the screw (whose design has been drastically improved since the original screw) and on the non-return valve at the screw tip.

In my opinion, the opening and closing movement of the screw tip is one of the few machine movements (if not the only one) that is not 100% controllable by the molder. As it moves the screw forward during filling, the machine causes the valve to move back through the outer ring, closing and sealing the material's passage. In the opposite metering movement, it is the pressure exerted by the material that advances through the screw and pushes the check ring forward, allowing the material to pass through to the front of the barrel .This opening and closing movement of the check ring is automatic, but not fully controlled by the molder. There are machine and screw manufacturers working on improving the design of this critical element of the process to provide better performance, durability and precision.

Performing the screw tip or check ring sealing study on a regular basis will ensure the correct operation of the machine, and therefore correct production. As explained in Section 3.4, two types of tests or studies can be carried out on this element's sealing, the dynamic test and the static test. Some of the problems that can occur with a check ring (whether worn, not sealed or in a poor condition) are: irregular cushion between shots, with no stability, irregular metering times, excessive part shrinkage, black spots burned into the material, streaks on the surface of the part, instability of part dimensions, and irregular part weight. This element of the injection unit should be carefully checked during scheduled maintenance for signs of wear or lack of sealing.

7.3.6.1 Dynamic Test Study

The weight of the molded part is an indicator of the sealing of the non-return valve or check ring. Depending on the sector or application of the molded part and the tolerances required in the parts, we must determine the maximum percentage of variability of the molded part's weight.

This test is carried out with 90–95% of the mold cavity filled.

Dynamic Screw Tip Sealing Study: Injection Machine Configuration

1. Decouple or disconnect the holding pressure phase

2. Set the holding pressure time and holding pressure to zero because the behavior of the screw tip sealing that we are going to test is mainly during the volumetric filling phase or dynamic phase, not during the holding pressure phase.

To carry out this study, we will perform 10 injection shots and control the weight of each of them to complete the test. For this test, we fill 90–95% of the cavities with the holding pressure phase set to zero (pressure and time) to analyze only the filling phase.

Start the process and once it has stabilized, collect 10 injection shots and check the shot weight obtained for each shot. If the screw tip is hermetic and sealed and there is no wear on the injection barrel, the weights are repeatable and consistent.

We can use the spreadsheet in Figure 7.54 to calculate statistical data, such as the maximum and minimum weight and the variability between shots, to check and maintain the precision or repeatability of the injection process. In the suggested spreadsheet, shot weight values such as average, median, range and variability are calculated. As a reference, variability of more than 3% indicates that the check ring valve should be replaced or disassembled and checked.

Figure 7.54 Example of a spreadsheet for a screw tip (check ring) sealing study

7.3.6.2 Static Test Study

This test is carried out with short screw movements.

Static Screw Tip Sealing Study: Injection Machine Configuration

Do not carry out this test on hot runner molds at high temperatures. This test or study involves the analysis and checking of the screw tip sealing or barrel wear.

1. Machine configuration for the static screw tip sealing study and the injection barrel wear study:

2. Allow the injection shot with its corresponding distribution channel or cold runner to cool down inside the mold so that there is no possibility of the material filling the mold

3. Set the holding pressure to a typical setting for the process, but at least 50% of the maximum holding pressure available on the setting machine

4. Set the holding pressure time to 10–20 seconds

5. Set the metering stroke to approximately 80% of the maximum screw volume

6. Set the switchover point based on stroke

7. Set the switchover point to change to the hold phase very close to the metering stroke position reached – the objective is to immediately change to the holding pressure phase

8. Inject a shot and check that the screw advances or remains stationary after the switchover point in the holding pressure stage

Keep in mind that the plastic will be molten and you should only see the screw advance slightly at the beginning of the injection movement, due to the compressibility of the molten polymer. After that, the screw must stop and remain static.

Carry out the same test with different metering levels or strokes or barrel capacity (i. e. with 10%, 25%, and 50% of the total available capacity). In other words, check at least three or four different metering strokes from the total metering stroke available on the machine setting. If the test at 80% capacity is acceptable, but the tests at 10% or 50% capacity are not, you may think that the screw tip sealing is correct and you may have wear on the barrel. If the screw advances slowly during testing at different metering positions, the screw tip should be checked or replaced immediately.

7.3.6.3 What about Hot Runners?

This static test study cannot be carried out on hot runners when at high temperatures. Bear in mind that one condition of the test is that no material be injected into the mold.

The test can be carried out on a cold runner. There are also companies that use injection nozzles with closed outlets to perform the static test. These kinds of nozzles allow us to activate the screw injection movement and perform a pressure injection without the risk of forcing material into the hot runner, since this material cannot flow through the nozzle because the material does not have an open outlet.

7.4 Scientific Injection Molding Tools Application Diagram

Below is a suggestion, accompanied by a diagram, for the sequential application of all these methods or studies, so that the whole process study is carried out in a logical sequence of application that allows us to take advantage of the different machine configurations to obtain the most information. Again, this is only a suggestion – it is not necessary to follow this sequence step by step to obtain the results; you can create your own sequence of studies relevant to you.

7.4.1 Logical Sequence of Application

7.4.1.1 Preliminary Studies and Calculations

- Estimation of the theoretical clamping force required (explained in detail in Section 7.1.1)

- Calculation of the maximum peripheral screw speed (explained in detail in Section 7.1.2)

- Calculation of the theoretical metering stroke or volume (explained in detail in Section 7.1.3)

- Estimation of the theoretical cooling time (explained in detail in Section 7.1.4)

7.4.1.2 Tools for Defining the Injection Molding Process

Subsequently, during the analysis or study of the process, using the tools suggested by the scientific injection molding methodology, we would follow this sequence:

- Relative viscosity or in-mold rheology or viscosity curve test (explained in detail in Section 7.2.1)

- Injection-speed linearity test (explained in detail in Section 7.2.2)

- Mold-filling balance study (explained in detail in Section 7.2.3)

- Optimal injection speed selection (explained in detail in Section 7.2.4)

- Mold filling study. Here we will also define the injection speed profile, if necessary, as well as define the switchover point (explained in detail in Section 7.2.5)

- Analysis of injection pressure losses along the filling system (explained in detail in Section 7.2.6)

- Delta P determination and study (explained in detail in Section 7.2.7)

- Gate sealing study (explained in detail in Section 7.2.8)

- Process window determination method (explained in detail in Section 7.2.9)

7.4.1.3 Further Tools for Checking and Improving the Injection Molding Process

Subsequently, we can complete the process definition with studies and calculations, such as:

- Shear rate at the gates study (explained in detail in Section 7.3.1)

- Cooling system study (explained in detail in Section 7.3.2)

- Cooling time study (for a dimensional characteristic) (explained in detail in Section 7.3.3)

- Residence time of the material in the injection unit study (explained in detail in Section 7.3.4)

- Filling time repeatability study – to check process stability (explained in detail in Section 7.3.5)

- Screw tip or check ring sealing study – dynamic test and static test (explained in detail in Section 7.3.6)

With all these tools or studies for the analysis and definition of the injection process, it is possible to define and monitor a process that will be robust, consistent and therefore productive – a process in which each important parameter, before being set in the machine control, has been previously contemplated and analyzed following this methodology.

7.4.2 Logical Sequence of Application Diagrams

7.4.2.1 Preliminary Studies and Calculations

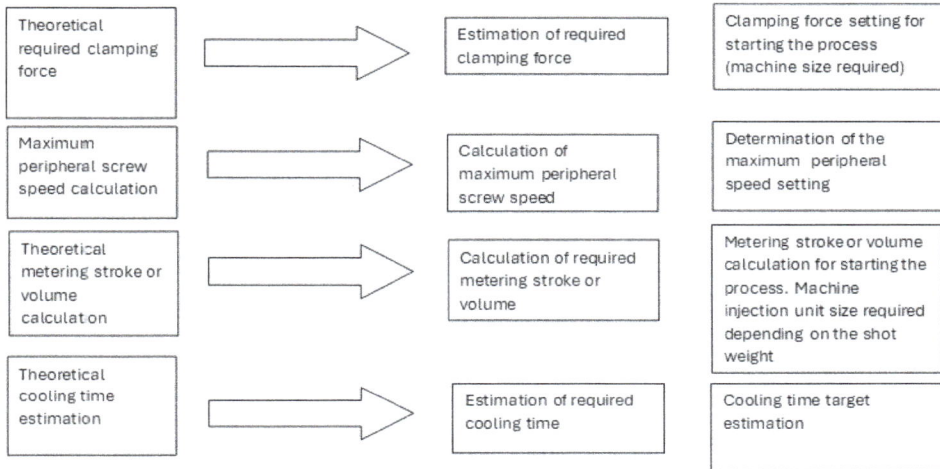

Theoretical required clamping force	→	Estimation of required clamping force	Clamping force setting for starting the process (machine size required)
Maximum peripheral screw speed calculation	→	Calculation of maximum peripheral screw speed	Determination of the maximum peripheral speed setting
Theoretical metering stroke or volume calculation	→	Calculation of required metering stroke or volume	Metering stroke or volume calculation for starting the process. Machine injection unit size required depending on the shot weight
Theoretical cooling time estimation	→	Estimation of required cooling time	Cooling time target estimation

7.4.2.2 Tools for Defining the Injection Molding Process

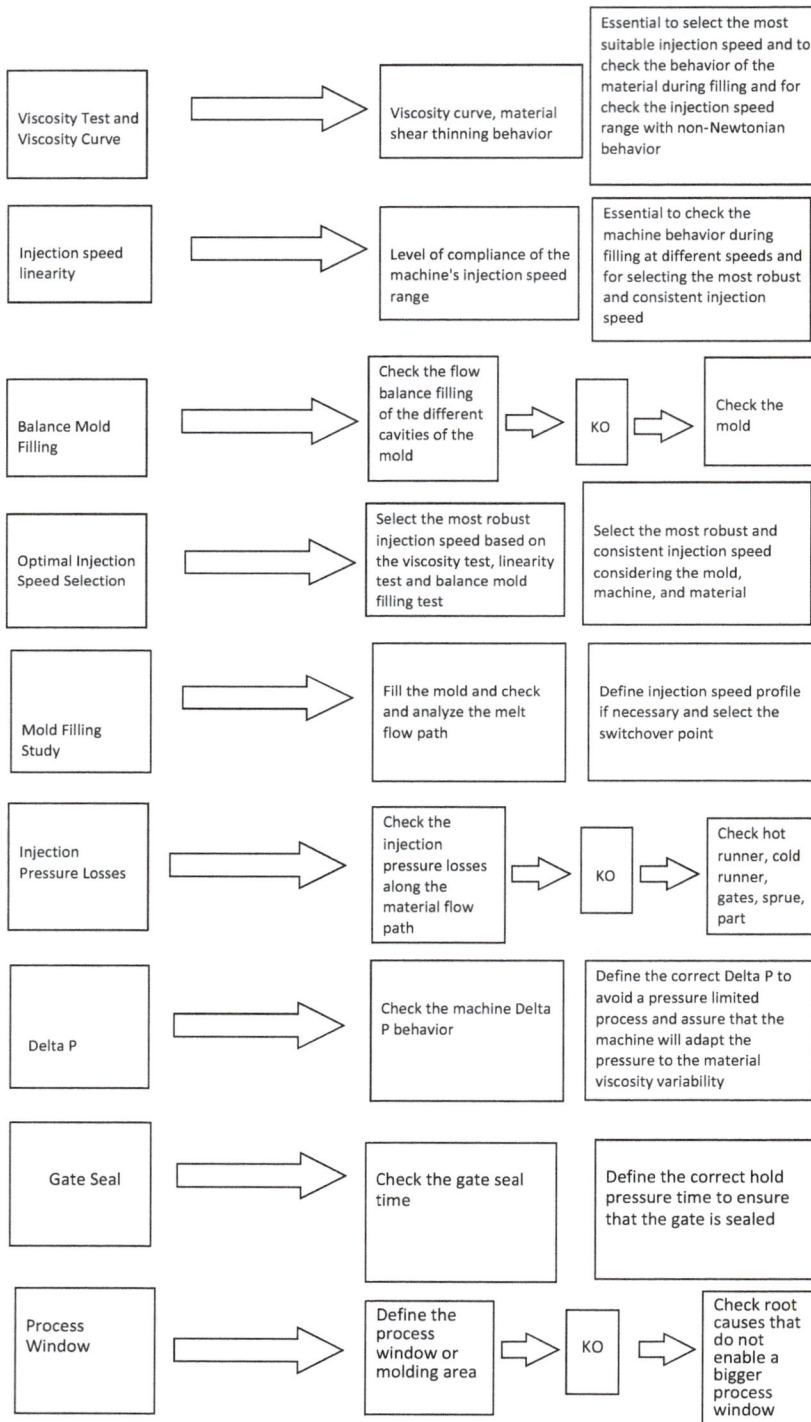

Viscosity Test and Viscosity Curve	→	Viscosity curve, material shear thinning behavior	Essential to select the most suitable injection speed and to check the behavior of the material during filling and for check the injection speed range with non-Newtonian behavior
Injection speed linearity	→	Level of compliance of the machine's injection speed range	Essential to check the machine behavior during filling at different speeds and for selecting the most robust and consistent injection speed
Balance Mold Filling	→	Check the flow balance filling of the different cavities of the mold	→ KO →	Check the mold
Optimal Injection Speed Selection	→	Select the most robust injection speed based on the viscosity test, linearity test and balance mold filling test	Select the most robust and consistent injection speed considering the mold, machine, and material
Mold Filling Study	→	Fill the mold and check and analyze the melt flow path	Define injection speed profile if necessary and select the switchover point
Injection Pressure Losses	→	Check the injection pressure losses along the material flow path	→ KO →	Check hot runner, cold runner, gates, sprue, part
Delta P	→	Check the machine Delta P behavior	Define the correct Delta P to avoid a pressure limited process and assure that the machine will adapt the pressure to the material viscosity variability
Gate Seal	→	Check the gate seal time	Define the correct hold pressure time to ensure that the gate is sealed
Process Window	→	Define the process window or molding area	→ KO →	Check root causes that do not enable a bigger process window

7.4.2.3 Further Tools for Checking and Improving the Injection Molding Process

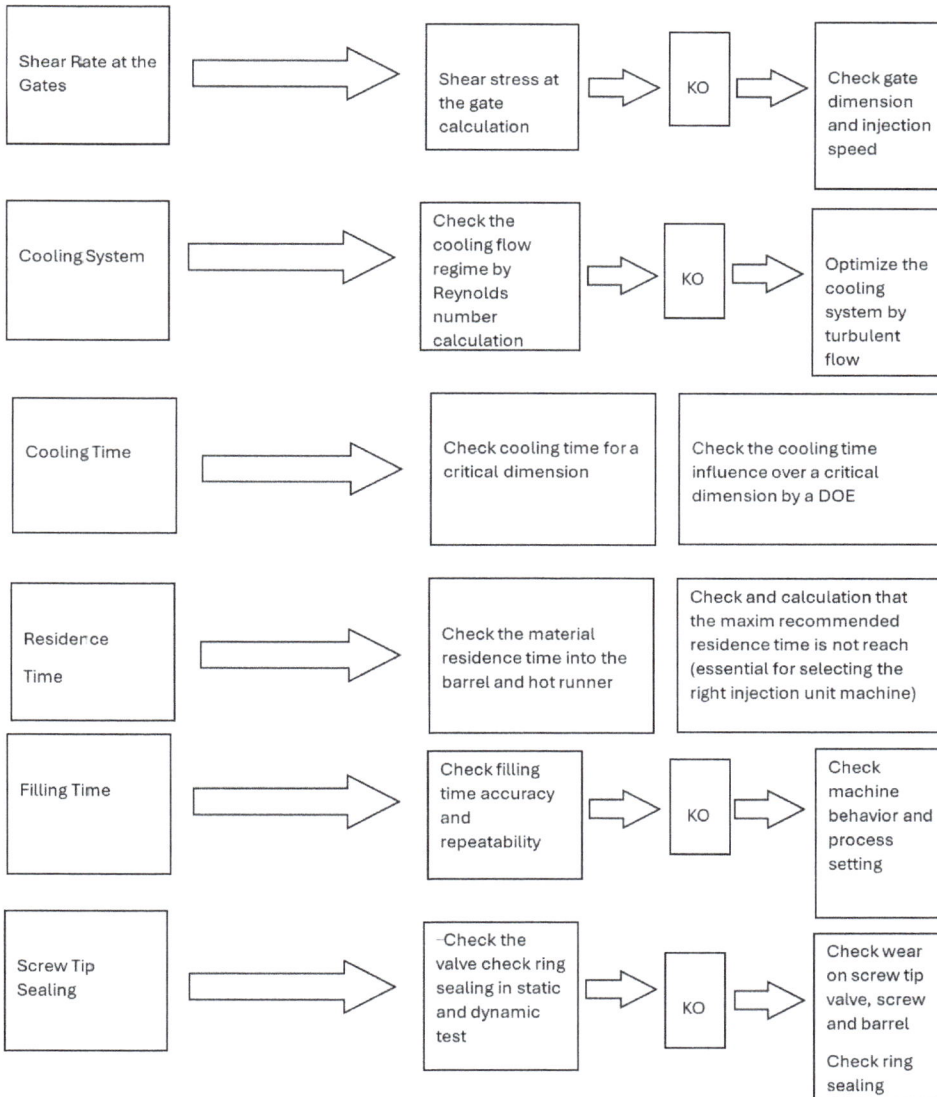

| Shear Rate at the Gates | ⟹ | Shear stress at the gate calculation | ⟹ | KO | ⟹ | Check gate dimension and injection speed |

| Cooling System | ⟹ | Check the cooling flow regime by Reynolds number calculation | ⟹ | KO | ⟹ | Optimize the cooling system by turbulent flow |

| Cooling Time | ⟹ | Check cooling time for a critical dimension | | Check the cooling time influence over a critical dimension by a DOE |

| Residence Time | ⟹ | Check the material residence time into the barrel and hot runner | | Check and calculation that the maxim recommended residence time is not reach (essential for selecting the right injection unit machine) |

| Filling Time | ⟹ | Check filling time accuracy and repeatability | ⟹ | KO | ⟹ | Check machine behavior and process setting |

| Screw Tip Sealing | ⟹ | Check the valve check ring sealing in static and dynamic test | ⟹ | KO | ⟹ | Check wear on screw tip valve, screw and barrel. Check ring sealing |

8 Top Ten Key Parameters in the Definition of the Injection Molding Process

Each of the parameters that make up a plastic injection molding process is important; however, we can highlight those of greatest importance along with the crucial inputs and outputs.

8.1 Injection Speed

The injection speed is the most critical speed in the injection process. The geometry of the part and the regulation of this speed determine if cavity filling will be fast or slow. A higher speed obviously shortens the injection or filling time and a lower speed prolongs it.

High injection speeds result in a rapid increase in injection pressure. In contrast, low injection speeds cause pressure drops in the nozzle and channels, due to the rapid growth of the solid layer. In this case, we lose the available section of the molten plastic flow channel and therefore cannot transmit the pressure far and optimally.

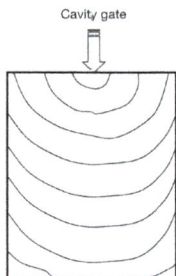

Cavity gate

Figure 8.1
Isochronous lines representing the progress of the flow front
(the lines should be as equidistant as possible)

The screw speed should control the progress of the material into the cavity. Filling should be as steady and fast as possible, so that the injection pressure gradient is constant. Figure 8.2 shows an ideal filling situation. Here, the flow front speed is constant and the injection pressure increases linearly.

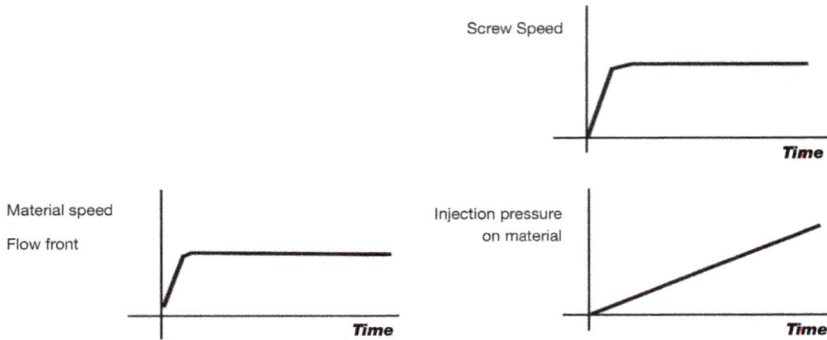

Figure 8.2 Ideal filling situation: constant screw speed along the injection stroke, constant material speed along the flow path during cavity filling, and proportional increase in injection pressure during filling

But molded parts usually have different sections and designs that facilitate acceleration or deceleration of the flow front during filling of the mold. The result is that the viscosity during filling of the cavity changes as a function of the flow front speed. If these speed changes are important, we can regulate the advance speed of the screw to obtain a more constant speed for the flow front.

Two typical situations for injection molding parts are shown below, namely when the flow front moves from a thin to a thick section and vice versa.

Table 8.1 Screw Speed Profile Influence over the Injection Pressure and the Material Flow Speed when Filling a Thin to Thick Cavity and, in Contrast, a Thick to Thin Cavity

	Let us imagine a part that is thin near the gate and thick at the end of the filling: thin to thick flow path.
	If our screw speed curve is flat...
	... the true speed of the flow front inside the cavity will decrease. This is because the volume available in the cavity is increasing and the volume of molten material entering the cavity is constant.

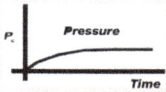

The hydraulic pressure will be flat because no increasing effort is required for meeting the set speed in the machine control.

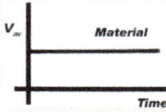

The flow front must advance into the cavity at as constant a speed as possible in order to obtain equidistant isochronous lines (see also Figure 8.1).

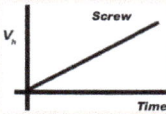

To achieve this, the speed set in the machine control must be increasing. We therefore require a higher screw speed in the filling area, which is thicker (i. e. more cubic centimeters of material per second to fill the cavity with the most available volume).

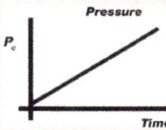

The hydraulic pressure will increase proportionally, since the machine's hydraulic system requires more pressure to maintain the set speed profile.

Let us imagine a part that is thick near the gate and thin at the end of the flow path: thick to thin flow path.

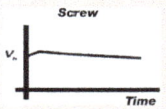

If our screw speed curve is flat...

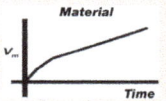

... the true speed of the flow front inside the cavity will increase. This is because the volume available in the cavity is decreasing while the volume of material input is constant.

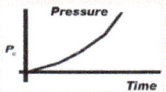

The hydraulic pressure required will rapidly increase disproportionately because the machine system is being put to the test in order to maintain the set speed in the machine control.

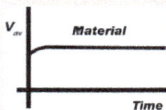

For proper cavity filling, the flow front must enter the cavity at as constant a speed as possible.

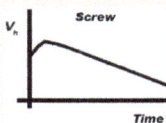

To achieve this, the set injection speed profile must be decreasing. The screw speed will therefore be lower when filling the thin part.

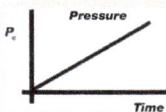

The hydraulic pressure will increase proportionally because the machine's pressure system will require proportionally more injection pressure to maintain the set speed in the machine control.
The planned decreasing speed avoids pressure peaks at the end of filling.

The key point here is that we must set a forward screw speed that makes it possible to maintain a constant flow front advance during cavity filling.

Filling speed

As a general rule, we must fill the cavity in the shortest possible time, not only to reduce the total cycle duration, but also to prevent premature expansion of the solid layer, which otherwise makes it difficult to properly pressurize the cavity. The rapid expansion of this cold solidified layer could be an important limitation in available for filling the cavity, resulting in significant pressure drops during filling. However, the following aspects must be taken into account:

- Very high speeds:
  - May cause overheating and thermal degradation of the material
  - May generate maximum shear stress in the material, causing breakage of molecular chains and consequent loss of properties
  - May cause stretching and displacement of the cold solidified layer – the inner molten layer will emerge outward, causing aesthetic defects and flaking
- Very low speeds:
  - May cause a reduction in material passage area, due to the increase in the cold layer in contact with the steel of the mold – this situation causes an increase in the injection pressure required to move the flow

What affects the filling speed? The filling speed is mainly determined by the following elements:

- Material
  - Flow rate, viscosity
  - Melt temperature
- Part design
  - Wall thickness
  - Thickness variations
  - Sharp corners, radii
  - Surface finish of mold and part
- Mold design
  - Gate sections
  - Runner passage sections
  - Mold heating and cooling system
  - Injection gate position
  - Venting, efficiency, and location

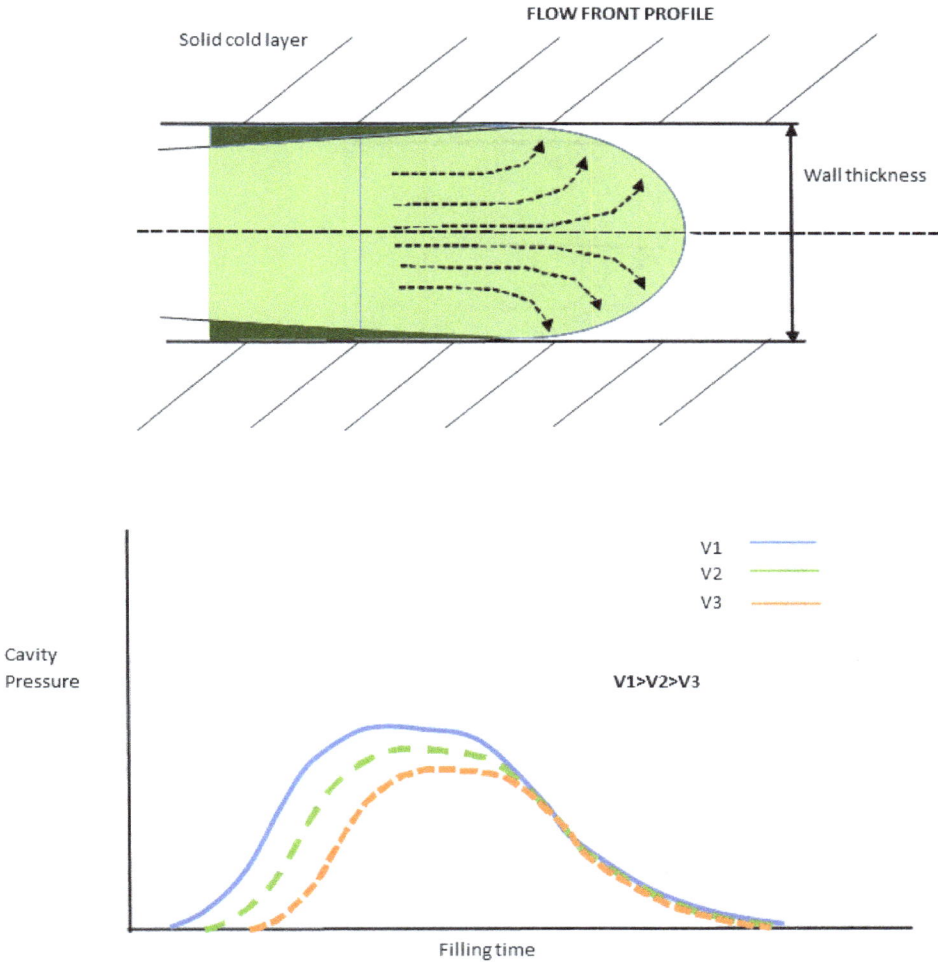

Figure 8.3 Flow front profile

The graphs in Figure 8.4 and Figure 8.5 show the strong influence of the filling speed on the injection and cavity pressures achieved. Increasing the filling speed results in the following:

- A reduction in the visibility of weld lines
- Greater mechanical strength of weld lines
- An increase in surface gloss of the part
- An increase in crystallinity
- An increase in melt temperature during filling
- An increase in the required clamping force

- A higher level of cavity pressure balance

- A higher level of surface orientation

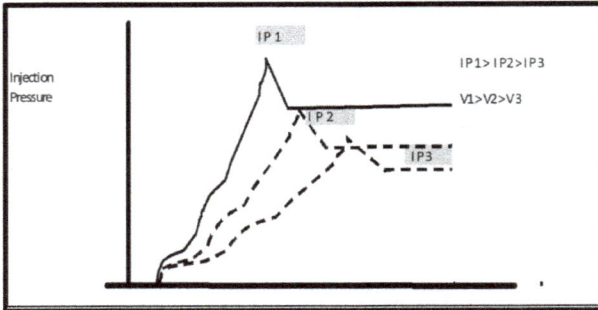

Figure 8.4
Injection pressure at different injection speeds

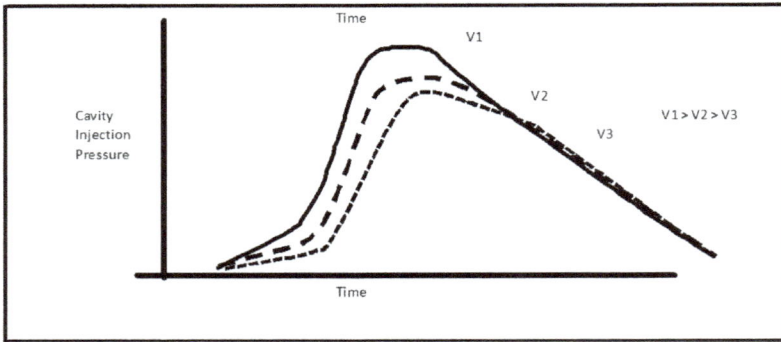

Figure 8.5 Cavity pressure at different injection speeds

8.2 Melt Temperature

This output parameter is the result of a combination of the following parameters or inputs:

- Peripheral screw speed

- Back pressure

- Injection unit temperature

The value obtained must be within the processing range recommended by the polymer manufacturer. The melt temperature selected will also depend on the following elements:

- Material viscosity
- Mold design (gate, hot runner, runners, section, cooling system, etc.)
- Part design (flow path length, thickness ratio, etc.)

8.2.1 Influence of the Melt Temperature

Increasing the melt temperature results in the following:

- A lower material viscosity and increased melt flow
- A reduction in molecular orientation
- A reduction in internal stress
- A reduction and improvement in weld line resistance
- Lower pressure drops in the mold
- An increase in shrinkage
- An increase in gas generation
- An increase in cooling time
- An increase in crystallinity
- An increase in surface brightness
- An increase in the tendency to form burrs and flash

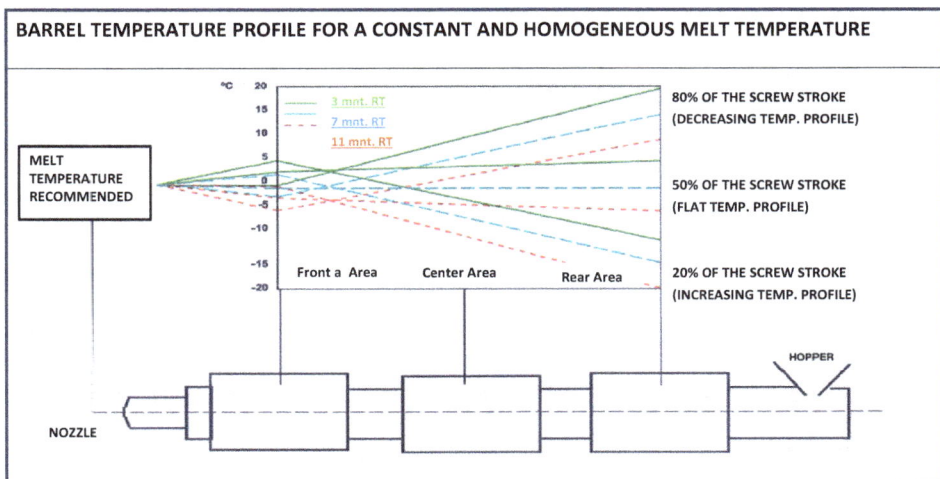

Figure 8.6 Barrel temperature profiles as a function of screw stroke utilization.
Source: DuPont

8.2.2 Residence Time

It is important to check that the maximum material residence time is not exceeded in order that critical material degradation may be avoided (see Section 7.3.4). The equation for calculating the residence time is:

$$\text{Residence time (min) RT} = \frac{\text{Resin weight into the barrel}}{\text{Shot weight}} \times \frac{\text{Cycle (s)}}{60}$$

Table 8.2 Amorphous Materials' Melt and Ejection Temperatures

Material	Melt Temperature (°C)	Ejection Temperature (°C)
PS	170–200	60
SB	180–280	90
SAN	200–260	110
ABS	200–270	100
PPO	250–290	200
Rigid PVC	170–210	50
Flexible PVC	140–200	60
PMMA	180–260	140
PC	280–320	140

Table 8.3 Semi-Crystalline Materials' Melt and Ejection Temperatures

Material	Melt Temperature (°C)	Ejection Part Temperature (°C)
PE-LD	190–200	80
PE-HD	210–300	110
PP	200–290	110
PA66	270–320	230
PA6	230–280	200
PA6 10	230–280	200
PA11	200–250	170
PA12	200–260	160
POM	190–220	150
PET	260–280	210
PBT	240–260	200

8.3 Peripheral Screw Speed

The melt temperature depends not only on the heat supplied by the electric barrel heaters, but also on the heat generated by friction and shear in the screw during melting and plasticization. The units used to determine the tangential screw speed are meters per second (m/s) and meters per minute (m/min).

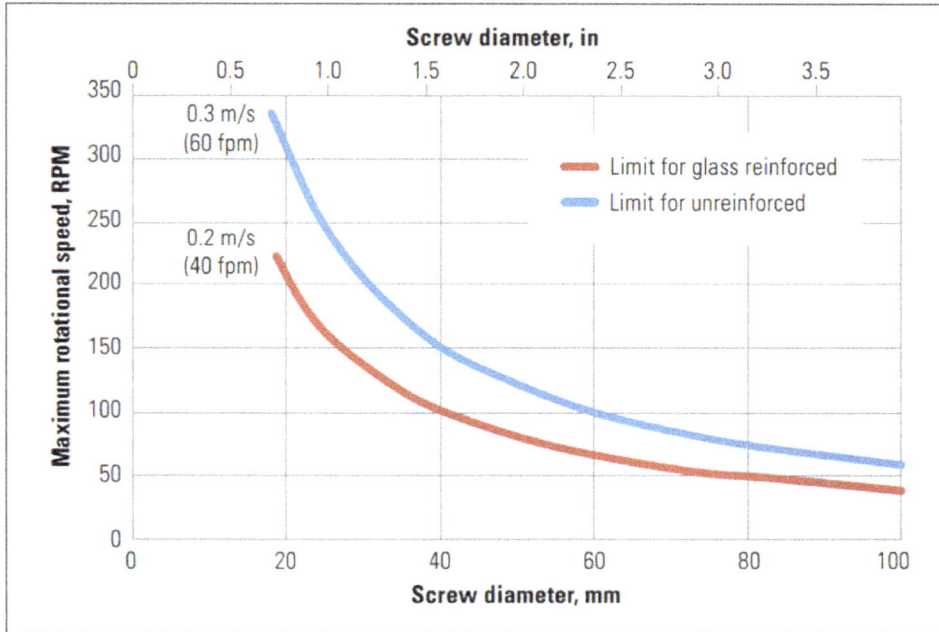

Figure 8.7 Tangential or peripheral speed as a function of screw diameter and rotation speed. Source: Eurecat

Table 8.4 Recommended Tangential or Peripheral Speed

Material	Tangential Speed (m/s)
PE	0.8
PP	0.7
PS	0.7
PA	0.5
POM	0.1 to 0.25
PET	0.3

Table 8.4 Recommended Tangential or Peripheral Speed *(continued)*

Material	Tangential Speed (m/s)
PBT	0.35
ABS, ASA	0.5
SAN	0.55
PC	0.5
PMMA	0.35
CA	0.45
PPE/PA -PPO blend	0.4
HYRTEL	0.4
ABS/PC	0.2
PA 66	0.8
TPU	0.2

8.4 Back Pressure

Back pressure is the effective pressure at the front tip of the screw during metering and metering. It is equal to the pressure exerted on the melt material by the rotation and design of the screw; the back pressure increases in the front area as the material is pumped from the screw to the front of the barrel, moving the screw backward until it reaches the metering position. This parameter is adjustable by the machine operator and can reach values of up to 100 bar of specific pressure.

This parameter can contribute to the proper thermal distribution and homogenization of the material, color masterbatch, additives, etc. and to better plasticization of the melt. The back pressure is very important when mixing pigments and additives, and we must take care when processing reinforced materials (i. e. with glass fibers, mineral fibers, steel fibers, carbon, etc.) to minimize the wear of the screw and prevent damage to the reinforcing fibers.

Effects of Back Pressure

Increasing the back pressure results in the following:

- More homogeneous melt
- More heat due to friction

- Displacement of air trapped in the pellets to the feed zone

- Reduction in metering variations and material cushion

These disadvantages must also be considered:

- Increased cycle time due to a longer metering time

- Sensitive materials may be thermally degraded

- Embedded reinforcing fibers may be damaged

- Increased abrasion and wear between the screw and the barrel

8.5 Injection Pressure

The injection pressure during the filling stage should be sufficient to maintain the injection speed set in the machine control and to obtain the desired filling time. Factors affecting the injection pressure are the same as those affecting the filling rate:

- Resistance to material flow

- Filling speed

- Hydraulic oil temperature

- Material temperature

- Mold temperature

The resulting filling pressure can be used to determine the correct settings for the factors that affect it.

Specific Injection Pressure

We must know the intensification ratio of the machines in order to always use and monitor the specific injection pressure or pressure applied to the material (Figure 8.8). This specific pressure is the one that tells us what pressure we are applying to the molten material and should be the same in any machine that we use to inject the same mold (see Section 9.1.3).

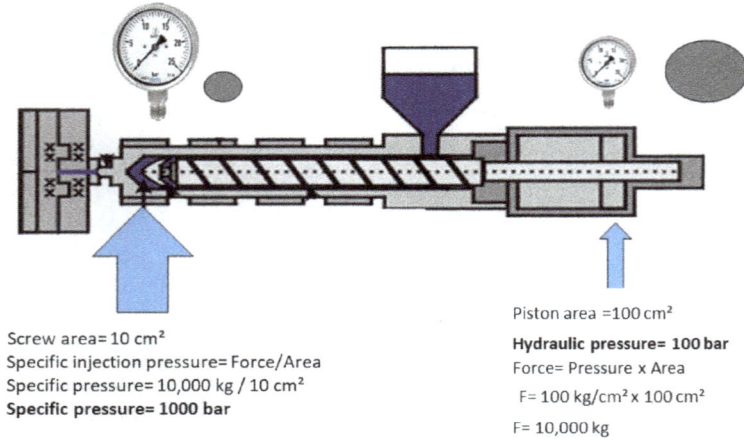

Screw area= 10 cm²
Specific injection pressure= Force/Area
Specific pressure= 10,000 kg / 10 cm²
Specific pressure= 1000 bar

Piston area =100 cm²
Hydraulic pressure= 100 bar
Force= Pressure x Area
  F= 100 kg/cm² x 100 cm²
  F= 10,000 kg

Figure 8.8 Intensification ratio diagram
100 bar hydraulic pressure = 1,000 bar specific pressure (i. e. an intensification ratio of 1:10).
According to the hydraulic and specific injection pressures chart, if we divide the specific pressure
by the hydraulic pressure, we know the intensification ratio between the piston and screw

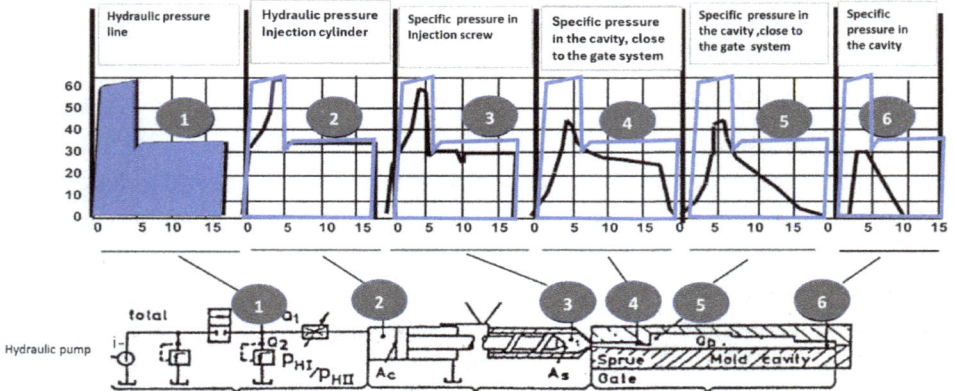

Figure 8.9 Pressure development along the hydraulic path (from the hydraulic oil
tank to the melt) – pressure drops occur for various reasons. Source: Eurecat

The injection pressure should be sufficient to maintain the set speed and, therefore,
the desired filling time. It is affected by the same factors as the injection speed.

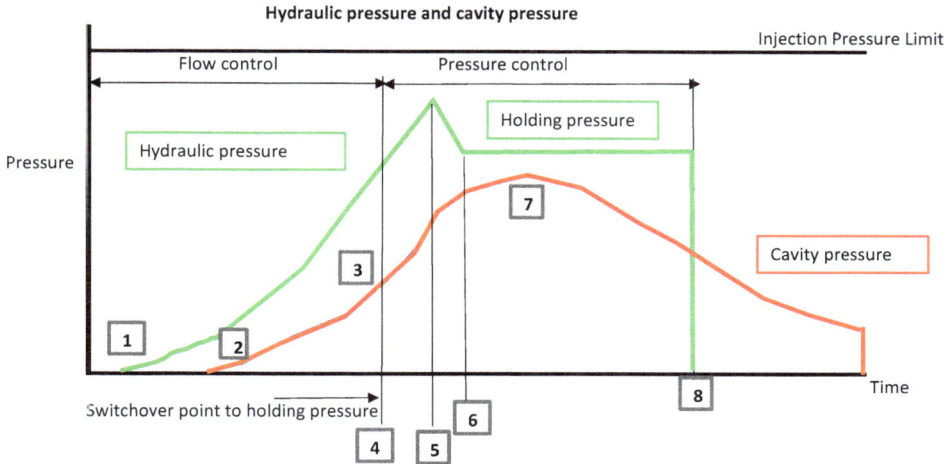

Figure 8.10 A graph showing hydraulic and cavity pressure throughout the injection process
1: screw starts moving forward; 2: cavity filling begins; 3: switchover point, electrical signal;
3–5: response time when switching to holding pressure; 5: maximum hydraulic pressure during filling; 6: holding pressure time control; 7: maximum pressure in cavity; 8: holding pressure time

8.6 Switchover Point

This is a critical part of the process – during volumetric filling of the cavity, we make the molten plastic enter the cavity at a certain speed. In this phase, the injection speed is controlled and the limitation is the available injection pressure limit.

Once the cavity is filled with melt, we must proceed to compensate for the loss of plastic volume that occurs when it begins to cool and the molecules rearrange themselves. In this phase, we compensate for the shrinkage or loss of volume by controlling the packing and holding phase, with the holding pressure and holding pressure time parameters used as controls during this phase.

The point for changing from the dynamic filling phase to the hold phase is a critical part of the process and is called the switchover point.

If we fill 100% of the cavity volume with molten plastic during the filling phase, this is a compressible material, like a rubber ball squeezed in our hand. Trying to fill 100% of the cavity volume with pressurized molten plastic is complicated and unpredictable because the search for this 100% fill may give rise to filling flash. It is recommended that the switchover point occur shortly before cavity volume filling reaches 100% (95–98%, depending on the material, injection speed applied, etc.).

Switching Systems in the Holding Pressure Stage

There are various systems for switching from the filling or dynamic phase to the packing and holding phase. The following is a summary and characteristics of these systems:

- Switching based on time
  - The worst of all systems
  - Ignores variations in viscosity
  - Ignores variations in melt temperature
  - Possible loss of precision when molding at high injection speeds

- Switching based on stroke or volume
  - The most commonly used system
  - Ignores possible material drop in the nozzle
  - Ignores transducer inaccuracies
  - Not recommended for high dose/capacity ratios
  - Ignores material viscosity variability

- Switching based on machine injection pressure
  - Reliable
  - Does not account for variations in viscosity

- Switching based on cavity pressure
  - Most reliable and expensive
  - Ignores melt and mold temperature variations
  - Compensates for variations in speed, viscosity, material dripping, etc.

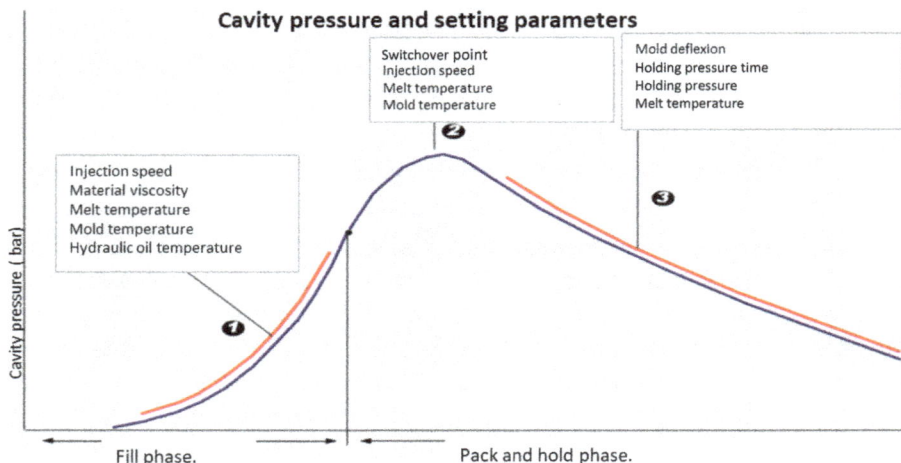

Cavity pressure and setting parameters

Switchover point
Injection speed
Melt temperature
Mold temperature

Mold deflexion
Holding pressure time
Holding pressure
Melt temperature

Injection speed
Material viscosity
Melt temperature
Mold temperature
Hydraulic oil temperature

Cavity pressure (bar)

Fill phase.   Pack and hold phase.

Figure 8.11 A graph showing how parameters affect cavity pressure

8.7 Holding Pressure

Filling, packing, and holding – three phases of the same stage:

After volumetric filling of the mold, it is imperative to pressurize the molten material that fills the cavity in order that the cavity aesthetic and dimensional characteristics of the part may be obtained. This is called the holding pressure phase. The complete filling of the mold during the holding phase can be divided into three sub-phases:

- Filling

  In this phase, the cavity is completely filled with molten plastic, but since plastic is a compressible material, we do not know when the cavity is completely filled. Remember that during dynamic filling, the mold should be filled to 95–98% volume to avoid overfilling the mold.

  The mold has now been filled to 100% volume.

- Packing

  In this phase, we must compensate for the volumetric shrinkage caused by the cooling effect and the molecular reorganization of the plastic. The goal is to compensate for the free volume that will be available, due to the volumetric material shrinkage or loss of polymer volume that will occur during cooling of the molten polymer (see Section 4.8.5 and Section 7.2.8.1). At this stage, the cavity is full of material, the goal is to pressurize the material inside the cavity. In this phase the maximum level of pressure in the cavity will be reached. It is in this phase that the part dimensions, weight, etc. will be defined.

  We control the pressure.

- Holding

  Once the phase 1 and 2 of volumetric compensation and pressurization or pack of the cavity is finished, it is necessary to correctly perform the third sub-phase of pressure maintenance or hold.

  In this phase, we must hold the pressure, not to force more molecules of molten plastic into the cavity, but to maintain the pressure until the gate freezes and closes, due to the cooling of the plastic. This prevents the plastic from flowing back out of the cavity – an effect called backflow. It can happen that the polymer inside the cavity subjected to pressures of hundreds of bar can flow backward and create a back flow, giving rise to a loss of properties and a loss of quality of the part. During this phase, we control pressure and time.

These three sub-phases are controlled by the values of holding pressure and holding-pressure time setting parameters.

Many molders do not differentiate between the packing and holding phases and set a value of pressure and time for the whole phase.

If volumetric compensation is not carried out correctly during the filling phase, this will result in internal voids (Figure 8.12).

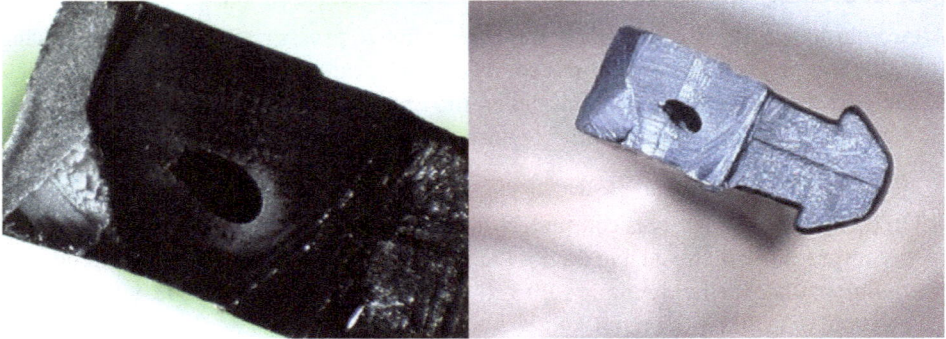

Figure 8.12 Examples of internal voids

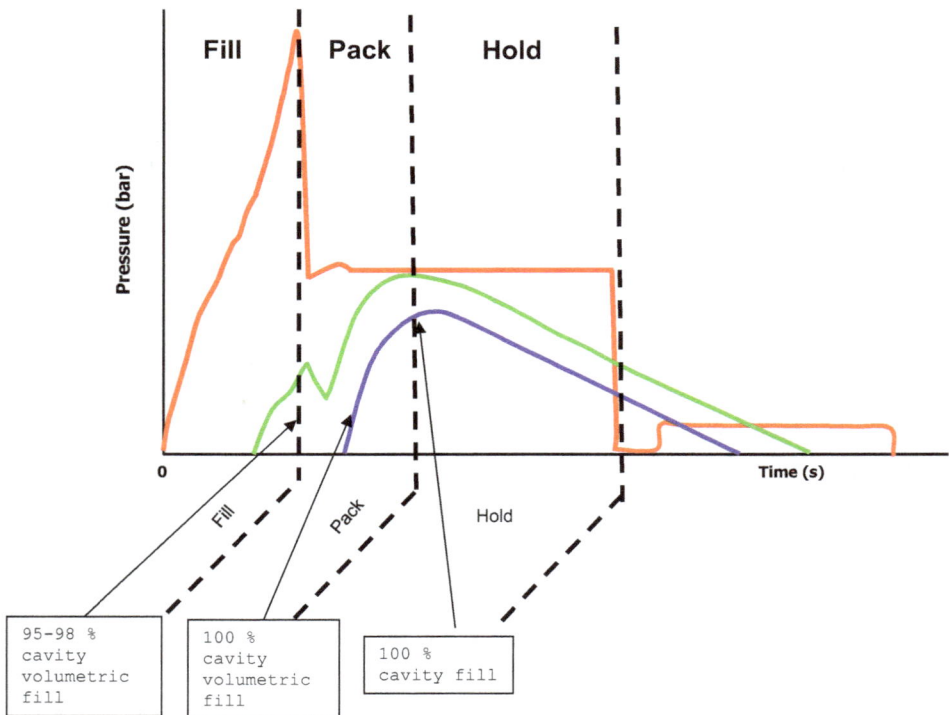

| 95–98 % cavity volumetric fill | 100 % cavity volumetric fill | 100 % cavity fill |

Figure 8.13 Injection pressure and cavity pressure during dynamic filling, holding pressure filling, packing and holding

The hold phase may affect the molded part in the following ways:

- Flash formation
- General shrinkage of the part
- Difficult part ejection
- Stress in the gate area
- Sink marks
- Internal voids
- Weld line resistance
- Part weight

8.7.1 Defining the Holding Pressure Time

The holding pressure time is an important parameter, so its correct definition must be approached with considerable care. Too short a holding time will result in an unstable and non-robust process. On the other hand, too long a holding time will result in long cycle times, productivity losses, stress near the gate, etc.

The holding pressure time can be determined by cavity pressure (in-cavity sensor required) and monitoring the molded part's weight.

- Cavity pressure drop: When this system determines the optimum holding pressure time, if the time is shorter than ideal, the cavity pressure will drop sharply when the machine pressure ceases. But if the time is long enough, the cavity pressure will remain and slowly decrease when the machine pressure ceases. This indicates that the gate has been sealed and the loss of pressure is due to the cooling of the material in the cavity (Figure 8.14).

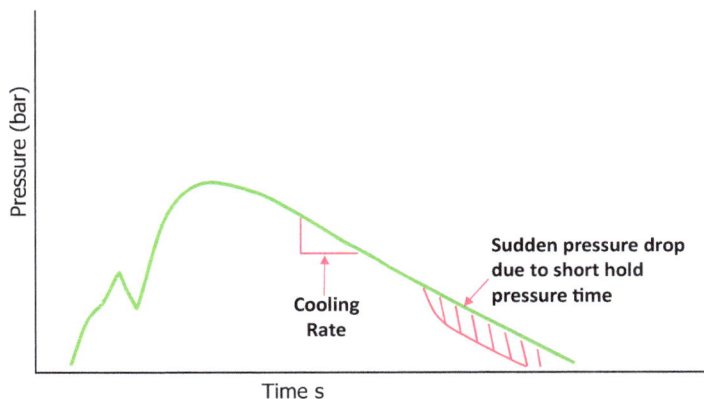

Figure 8.14 Cavity pressure development during the hold phase and cooling

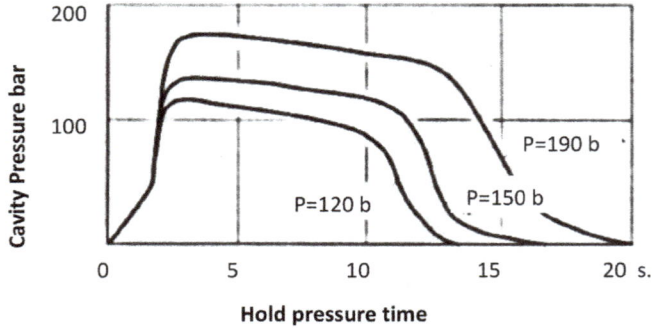

Figure 8.15 Cavity pressure with different holding pressures

- Monitoring the molded part's weight: The optimum time provides us with the maximum weight of the part. Increasing the holding pressure time increases the weight, due to more material being pushed into the cavity. However, there is a moment when the weight stops increasing despite the time continuing to increase, because the gate is frozen and closed. When controlling by weight, runners must be ignored.

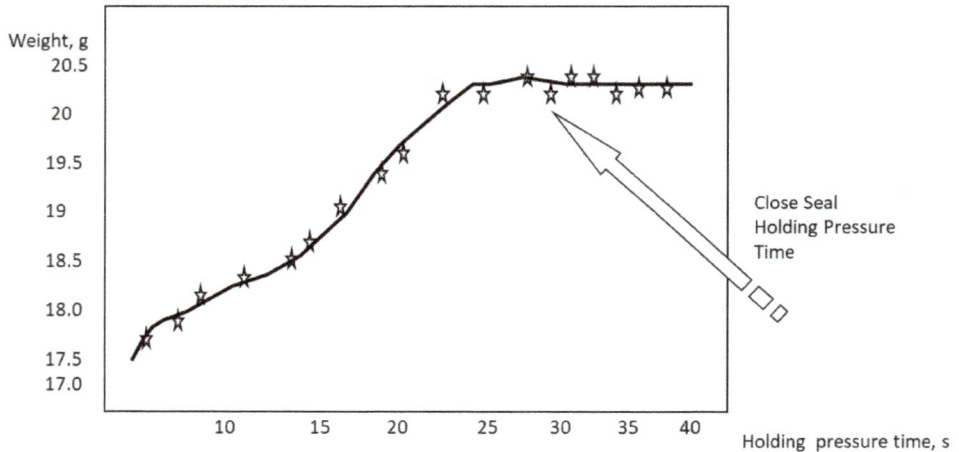

Figure 8.16 Molded part weight development for determining the optimum holding pressure time

8.7.2 Sealed or Open Gate?

As explained in Section 7.2.8.5, the choice between a sealed or an open gate depends on the part, application, requirements, etc. A sealed gate provides repeatability of process and dimensions, but also possible stress in areas near the gate; an open gate brings a greater possible distribution of measurements, machine precision is essential (time-dependent process), less stress near the gate, less pressure gradient = less distortion or warping, and possible backflow. It is recommended to mechanically test both options.

8.8 Mold Temperature

Mold and cavity temperatures determine the cycle time and the quality of the molded part's internal structure. Low mold temperatures result in shorter cooling times and higher cooling speeds, which can adversely affect part quality.

For semi-crystalline thermoplastics, the properties of molded parts depend on the cooling speed. Rapid cooling leads to an amorphous, thick outer layer and low crystallinity. On the other hand, slow cooling results in high crystallinity, stable lamellae and crystals, and better mechanical properties.

The ideal situation is homogeneous cooling resulting from a homogeneous temperature distribution in the mold and in the melt. To achieve this, correct thermal conditioning of the mold is necessary.

The material we are molding determines the actions we must act take to achieve an appropriate mold temperature:

- Cooling with water
  - Temperature control with pressurized water up to temperatures of 150–180 °C
  - Temperature control with oil to temperatures above 150 °C
  - Heating with electrical heaters for higher temperatures
- Increasing the mold temperature, which leads to:
  - Increased surface brightness (gloss)
  - Reduced internal stress
  - Increased impact strength
  - Improved weld line resistance and appearance
  - Increased mold shrinkage
  - Increased flash formation
  - Increased cooling time

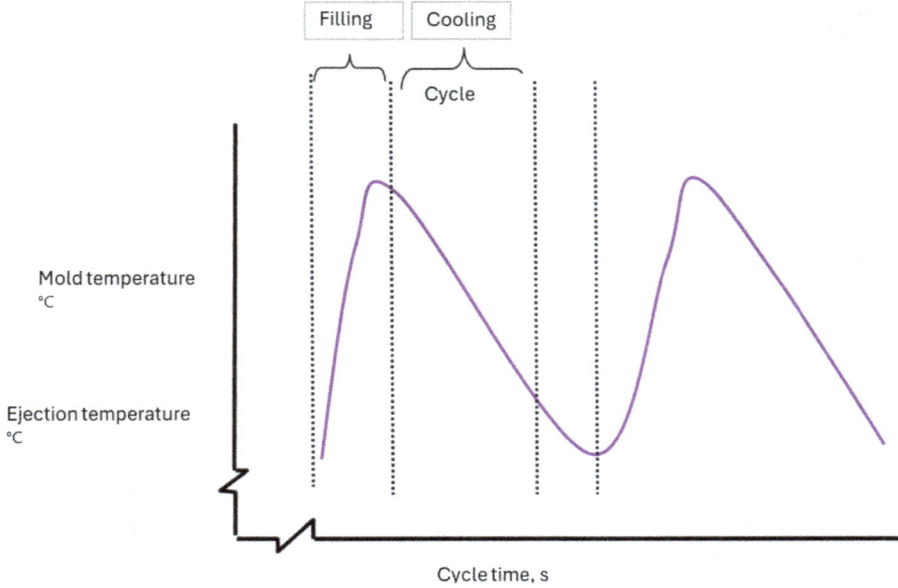

Figure 8.17 Temperature behavior in the mold wall at each injection cycle

Mold temperature affects the molded part in various ways:

- Mold shrinkage and post-shrinkage
- Surface gloss
- Internal stress
- Impact strength
- Weld lines
- Flash
- Injection time

Table 8.5 Mold Temperatures Recommended for Some Materials

Amorphous	Mold Temp. (°C)	Semi-Crystalline	Mold Temp. (°C)
PS	20–80	PE-LD	20–60
SB	10–60	PE-HD	20–60
SAN	40–80	PA6	80–90
ABS	60–80	PA66	80–90
PVC	20–60	PA610	40–90

Amorphous	Mold Temp. (°C)	Semi-Crystalline	Mold Temp. (°C)
CA	50–80	PA12	40–80
CAB	50–80	POM	40–120
PMMA	40–80	PAT	90–160
PC	80–120	PBT	40–100
PPO	80–120	PPS	130–150
PA	70–100		

8.9 Metering

Experimentally, the optimum screw stroke has been determined to be between one and three screw diameters equivalent. Metering strokes of less than one diameter and greater than four diameters should be avoided. See Section 2.6 and, particularly, Figure 2.18.

The volume used in an injection unit is approximately 70–80% of its maximum capacity and a minimum of 20–30%.

8.10 Cushion

Cushion is the residual volume remaining at the front of the screw at the end of the holding pressure stage. It is essential to have some cushion to ensure the proper application and "upstream" transmission of the injection pressure. It also absorbs volume differences between cycles, providing stability in the weights and volumes injected.

As a general rule of thumb, the cushion should be about 10% of the metered amount.

Spring effect with large cushion:

Due to the compressibility of plastics, a large cushion does not transmit the compression pressure as well as a smaller cushion. The cushion acts like a spring made of molten material; when pressure is applied during the filling or dynamic phase, and also when the holding pressure is applied, this molten material absorbs part of the pressure exerted by the machine and does not transmit it completely to the mold. The larger the cushion, the worse the pressure transmission.

Oven effect:

As the material moves from the hopper to the front of the injection unit, it receives heat from the rotation and compression of the screw. Once the material is in the front of the barrel, it no longer receives heat from the movement of the screw, but only from the external electric heaters; consequently during all the time it remains in this zone (cushion), the material is heated externally, with the result that there is overheating of the material in contact with the barrel's internal diameter.

This parameter provides very good information about the stability, consistency and precision of the process. In general, and depending on the screw diameter, the cushion should have a volume equivalent to 5–10% of the injected volume.

8.11 Other Key Aspects

8.11.1 Melt Preparation

Melt preparation and optimization: It is essential to prepare a proper melt in terms of temperature homogeneity, density repeatability, and metering volume repeatability (see Chapter 10). Without a high-quality melt, the injection process cannot be repeatable and consistent.

To achieve this, we need to monitor and regulate the parameters that affect the quality of the melt. Some of them have already been discussed in this chapter, but others below are worth mentioning:

- Temperature profile of the injection unit (see Section 8.2)
- Throat and tracking temperature
- Screw rotation speed – tangential/peripheral speed (see Section 8.3)
- Back pressure
- Decompression stroke and speed
- Metering, metering stroke / screw diameter ratio (see Section 8.9)

8.11.1.1 Throat and Tracking Temperature

In this zone, the coefficient of friction between the material, the screw and the barrel must be correct; this coefficient of friction must favor our intention, which is to efficiently advance the material to the front of the injection unit. To achieve this, the coefficient of friction of the material with the barrel must be greater than that of the material with the screw.

Advantages of correct tracking temperature:

- Improves material filling inside the screw

- Prevents condensation

- Reduces the temperature difference between the front part of the screw and barrel and the material inlet

- Improves the performance and durability of the rear heaters

- Reduces slippage of the solid pellets in the screw by generating the right level of friction between the material and the barrel's inner surface

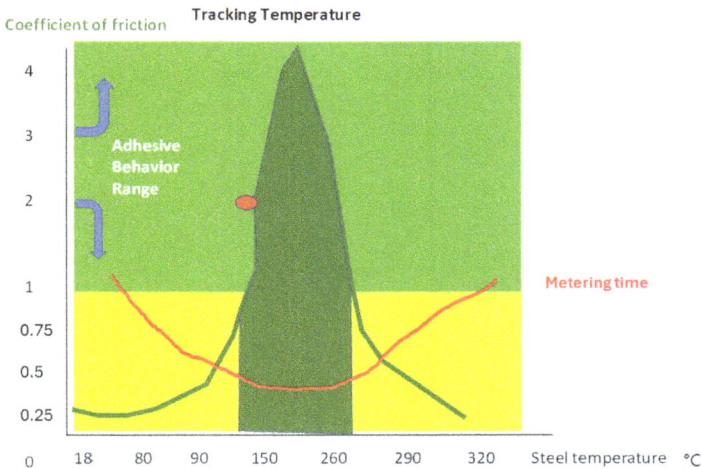

Figure 8.18 Coefficient of friction and metering time development at different throat steel temperatures

8.11.1.2 Decompression Stroke and Speed

The decompression movement, especially the one after metering, aims not only to relieve the pressure on the material in the front part of the barrel, but also to bring the sealing ring of the screw valve to the foremost front position. This ensures that the closing stroke of this ring will always remain constant and, therefore, possible leakage of material during the closing movement of the ring will also remain constant. As for the decompression movement speed, it must be fast enough to ensure the position of the sealing ring, but it must also enable the position reached by the screw, and therefore by the sealing ring, to be repeated.

To determine both the stroke and speed of the decompression movement, a series of DOEs (Design of Experiments) can be conducted, controlling the total shot weight of each cycle with different stroke and speed settings. The optimum settings will be those which provide the smallest shot weight range (see Section 10.6).

8.11.1.3 Back Pressure

Back pressure has several purposes. Firstly, it provides energy to homogenize and plasticize the material as it passes through the injection unit. It prevents the air in the pellets from moving forward to the front of the barrel. By exerting pressure on the molten material in the front of the barrel, the back pressure directly affects the density of the molten material and therefore the weight of the shot. Maximum regularity in this density is essential for achieving injection cycles with regular and consistent weights.

To determine the optimum back pressure, a series of DOEs can be conducted, controlling the total shot weight of each cycle with different back pressure settings. The optimum settings will be those which provide the smallest shot weight range (see Figure 8.19).

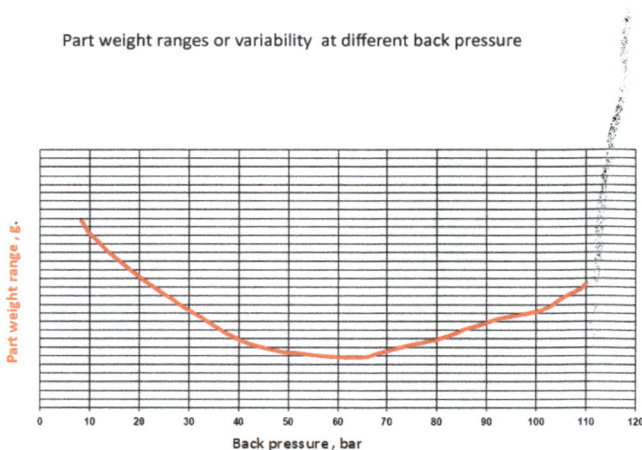

Figure 8.19 A graph showing optimized screw back pressures

8.11.2 Machine Input Settings and Process Outputs

As introduced in Section 1.2.4, one of the most important paradigm shifts that injection molding technicians need to make is the shift from understanding injection molding from a machine perspective to a mold and material perspective.

8.11.2.1 Parameter Data Sheets

Most injection molders use parameter data sheets and document the machine parameter settings, but not the material data. What is the difference?

When we establish a process and implement it, it provides us with very important information that indicates how the process is working and how the different variables involved in the process are interrelated. This information received from the process is a consequence of the machine setting parameters or the process setting, as well as the rest of the variables that affect the process. These values are independent of the machine input settings – they are the process outputs. This information is essential for proper control over the injection process. For more advanced injection methods, it is vital to use the process outputs both to define processes and to analyze the causes of possible process deviations.

As important and useful as parameter data sheets and machine setting records are, they are only inputs to be entered into the control system of the injection molding machine. In advanced injection molding, the focus is also on process outputs. If we want the process to be repeatable, batch after batch, there is no alternative – we have to record, check, and control the process outputs; machine settings do not tell us what is really happening in the process.

It seems logical that if we repeat the machine setting parameters from one batch to the next on the same machine, the machine process parameters will be the same and the parts produced will be identical. However, this is not always the case.

The recorded machine setting parameters allow us to establish a starting point. They are necessary, but they neither describe nor identify what is happening in the process, in the plastic and in the mold we are using to mold the part. This is very common when molds are transferred between injection molding plants, or even more simply, when parts are manufactured on different machines in one injection molding plant, or even when different production batches are molded on the same machine. How many times have we heard injection molding technicians say, "All the machine setting parameters are the same as always". Nothing seems to have changed in the settings, but if there has been a change in the parts, there has most likely been a change in the process outputs.

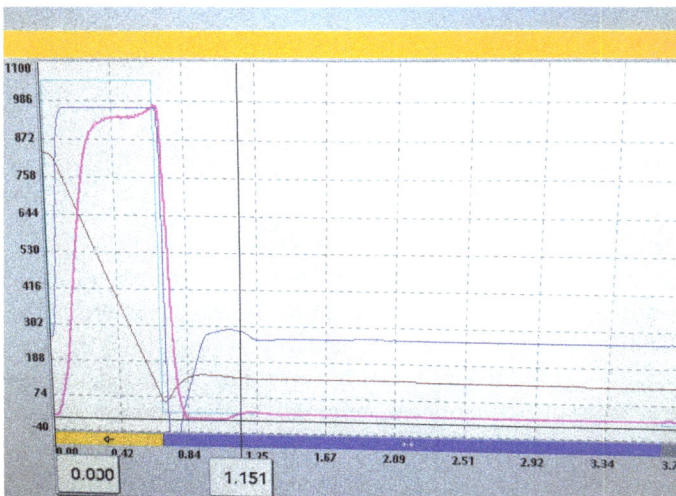

In plastic injection molding, the focus is typically on machine-dependent input values. For example:

- Injection unit and hot runner temperatures

- Injection speed setting

- Multiple time settings (holding pressure, cooling time, etc.)

- Multiple pressure settings (injection, back pressure, holding pressure, etc.)

- Clamping force setting

- Screw rotation speed (rpm)

8.11.2.2 Process Outputs

Process outputs are defined as the data received through the process as a result of the process and its various interacting variables. Outputs are not data that can be programmed directly by the molder.

Process outputs should be recorded. If we group the possible outputs of interest, we might establish the following classification (see the more detailed list in Section 1.2.4):

- Temperatures

- Times

- Pressures

- Weights

- Additional data

  - In addition, the advanced molding technician can document and add outputs through the scientific injection molding tools and methodology, such as:

- Residence time

- Peripheral screw speed

- Screw intensification ratio

- Dose/diameter ratio

- Viscosity curve

- Etc.

The process outputs, studies and calculations greatly enhance the process information and help to make the right decisions when defining the process parameters. The process documentation should include all outputs considered necessary for the correct recording of the process for it to be subsequently controlled. Machine settings alone are not enough; we are missing valuable and essential information for process control and improvement.

8.11.3 Process Tolerances

On many occasions, we ask ourselves what percentage of variability is acceptable for each key process parameter. The answer is not simple, considering that not all industries or applications have the same requirements, and that setting maximum and minimum tolerances for some key process parameters can narrow the process window to a point where it becomes completely unproductive.

The correct tolerances must be determined through process testing, by carrying out a process window study, e. g. for certain injection molding parameters, or by using a DOE with different factors to determine real maximum and minimum tolerance limits for certain parameters.

It can also be helpful to apply a generic percentage of tolerance for some key parameters, taking into account the type of part to be molded, such as the values shown in Table 8.6. However, bear in mind that these are generic percentages, not specific to any particular mold, part, machine, or material.

Table 8.6 Parameter Deviations by Type of Part Molded (Expressed as %)

Parameter	Precision Parts	Technical Parts	Commodity Parts
Metering time	1	2	5
Injection time	0.5	1	2
Switchover point to holding pressure	1	1.5	2
Cushion	2	3	4
Melt temperature	1	2	3
Clamping force	2	3	4
Back pressure	5	8	10
Metering stroke	0.1	0.2	0.3

9 Process Portability, DOE (Design of Experiments), Mold Qualification and Process Validation

9.1 Process Portability

In the plastic injection molding industry, there are a few thousand active companies. If we look at all of them, we will find very well-prepared and structured process parameter sheets, with a lot of information that allows injection molding technicians to properly record the process conditions.

We sometimes apply the same parameter settings recorded in the parameter sheets to the same machine (and even worse, on different machines), but the result, the production or the quality of the parts is not the same as in the homologation or the previous production batch. Then we look at the process, we compare it with the setting parameters recorded in the parameter sheet and, surprisingly, we find that "everything is the same" yet we do not get the same part quality.

Injection molding facilities invest a lot of time and resources in the launch and validation of a new project that introduces new molds to the molding plant by carrying out the necessary mold tests, retouching, homologation tests, run and rate, capacity studies and FMEA, amongst many others, for the purpose of parts homologation and mold and process validation. The intention is that once the product, the mold and the process have been approved, they will be repeatable in subsequent series under the same approval conditions, on the same machine and with the same setting parameters, until the end of the project. In most cases, once the process and the product have been approved, the intention is that they should also be repeatable on other machines with different characteristics from the machine used in the approval (i. e. the goal is to always produce correct parts on the same machine in different production runs and even on similar machines from the facilities' fleet, but this does not always occur).

Nowadays, production batches are becoming shorter and shorter, there are more and more mold changes every day, and we have to stop and start the process many times during the product's lifetime; we also have to manufacture the same parts in increasingly greater quantities, in exactly the same quality, on different machines than those available in the fleet, not to mention the production transfers that take place between plants in different geographical locations. We can therefore conclude that it is now essential to be able to use different machines that have different characteristics to manufacture parts in the same mold, or to be able to repeat the process several times on the same machine, manufacturing smaller and smaller batches, which is quite a challenge.

This difficulty of repeating the process, on the same machine or on another machine, is minimized by using a good portability system, as otherwise we will be inviting "Murphy" to the party, and he will come with all his friends.

9.1.1 Portability

When we change the mold and we start the production process again, there are several variables that can affect the proper repeatability of the process. There are two things to keep in mind:

- We should record all the outputs when the correct process is stabilized and running fine.

- We should move or convert inputs and outputs from the correct process to any machine with similar or different characteristics – this is called portability.

Portability is the process data collection strategy and the scientific or mathematical ability to reproduce a process on different machines while maintaining the original process conditions.

It is essential to be able to use different machines that have different characteristics to produce the same mold and parts. It is also essential to repeat a correct process on the same machine to produce identical parts in different production runs, batches, etc.

The difficulty of repeating and reproducing the original process on different machines is minimized by good portability. To achieve this, it is necessary to:

- Always use specific units for each magnitude or parameter

- Know the exact injection machine characteristics (screw diameter, intensification ratio, L/D, maximum metering stroke, clamping force (t), Delta P, etc.)

- Collect the essential outputs from the original process

- Apply calculation systems or parameter conversions with different injection machines

9.1.1.1 The Specific Units

The specific units are "wildcard" units, because when using them, we are working with units that can be applied to any machine, regardless of its own machine characteristics. They are units that will help us to set and compare the process conditions of different machines with the same parameters set. Typical units specific to the injection process are shown in Table 9.1.

Table 9.1 Parameters and Their Specific Units

Parameter	Specific Unit
Metering	cm^3
Injection filling rate	cm^3/s
Injection filling time	s
Holding pressure	bar
Injection filling pressure	bar
Cushion	cm^3
Switchover point	cm^3
Pressure at switchover point	bar
Peripheral screw speed	m/s; m/min
Back pressure	bar
Injection speed profile switchover points	cm^3
Cycle time	s
Metering time	s
Cooling time	s
Melt temperature	°C
Mold temperature	°C

If we know all this information (parameters and outputs in the correct specific units), we can convert them into the same parameters and outputs of the original process and replicate them exactly on different injection molding machines with different characteristics.

Figure 9.1 is an example of a spreadsheet for the conversion from the original process parameters to the parameters of another injection molding machine with different characteristics. Since this conversion is a mathematical calculation, it will be necessary to take into account possible deviations due to the wear, inertia, different acceleration, etc. of each machine.

SCIENTIFIC MOLDING			CALCULATIONS AND PORTABILITY			Put data in green cells	
PART NAME			MATERIAL	ABS		DATE	
MOLD NUMBER			MELT DENSITY	0.95	g/ cm 3		
MACHINE A	150	Tn	MACHINE B	480	Tn		
SCREW DIAMETER	50	mm	SCREW DIAMETER	70	mm		
DOSAGE STROKE	115	mm	DOSAGE STROKE	54.70	mm		
DOSAGE VOLUME	225.8	cm³	DOSAGE VOLUME	225.80	cm³		
WEIGHT	200	g	WEIGHT	200	g		
MAXIMUM DOSAGE STROKE	300	mm	MAXIMUM DOSAGE STROKE	500	mm		
RESIDENCE TIME	2.81	minutes	RESIDENCE TIME	7.68	minutes		
DOSAGE: SCREW DIAM RATIO	2.30	times	DOSAGE: SCREW DIAM RATIO	0.78	times		
INJECTION UNIT UTILIZATION (%)	38.3	%	INJECTION UNIT UTILIZATION (%)	10.9	%		
CYCLE TIME	45	s	CYCLE TIME	45	s		
FILLING TIME	2	s	FILLING TIME	2	s		
REAL INJECTION LINEAL SPEED	55	mm/ s	REAL INJECTION LINEAL SPEED	28.06	mm/ s		
REAL INJECTION VOLUME SPEED	107.99	cm³ / s	REAL INJECTION VOLUME SPEED	107.99	cm³ / s		
INTENSIFICATION RATIO	10	:1	INTENSIFICATION RATIO	8	:1		
SPECIFIC INJECTION PRESSURE	1400	bar	SPECIFIC INJECTION PRESSURE	1400	bar		
HYDRAULIC INJECTION PRESSURE	140	bar	HYDRAULIC INJECTION PRESSURE	175	bar		
BACK PRESSURE	10	bar	BACK PRESSURE	12.5	bar		
FILLING CAVITY STROKE	110	mm	FILLING CAVITY STROKE	56.12	mm		
FILLING CAVITY VOLUME	216.0	cm3	FILLING CAVITY VOLUME	216.0	cm3		
SCREW RPM	80	rpm	SCREW RPM	57.14	rpm		
SCREW PERIPHERAL SPEED	0.21	m / s	SCREW PERIPHERAL SPEED	0.21	m/ s		
COOLING TIME	15	s	COOLING TIME	15	sc		

Figure 9.1 Example of a spreadsheet for portability calculations. Parameters from one injection machine's original process are converted into the parameters of another injection machine

9.1.1.2 The Essential Outputs

In order to achieve proper process portability and to be able to reproduce a process on different injection molding machines, it is essential to collect a set of outputs from the original process. These will be essential for reproducing the process and changing the molding conditions in the event of process deviations. All the essential outputs should be managed in specific units; see Table 9.2.

Table 9.2 Process Outputs and Their Specific Units

Output	Specific Units
Melt temperature	°C
Back pressure	bar
Peripheral screw speed	m/s; m/min
Metering time	s
Filling time	s
Part weight at switchover points	g

Output	Specific Units
Filling maximum pressure peak	bar
Purge injection pressure	bar
Holding pressure	bar
Holding pressure time	bar
Cavity maximum pressure (if available)	bar
Part weight	g
Total shot weight	g
Gate seal time	s
Cooling flow rate	l/min
Coolant in temperature	°C
Coolant out temperature	°C
Cycle time	s
Residence time	s
Delta P	bar
Cushion	cm^3
Back pressure	bar
Throat temperature	°C
Cavity steel temperature	°C

Note that all pressures in Table 9.1 and Table 9.2 are in specific injection pressure or pressure applied to the melt material.

9.1.1.3 Important Process Inputs or Settings

We must not also forget the importance of process inputs or settings. While all settings are important, some of them must not be absent from any of the parameter sheets; see Table 9.3.

Table 9.3 Process Inputs or Settings and Their Specific Units

Input or Settings	Specific Units
Barrel temperature profile	°C
Cooling time	s

Table 9.3 Process Inputs or Settings and Their Specific Units *(continued)*

Input or Settings	Specific Units
Decompression	cm^3
Decompression speed	cm^3/s
Back pressure	bar
Injection filling speed profile	cm^3/s
Injection filling speed profile switchover points	cm^3
Switchover point	cm^3
Holding pressure	bar
Holding pressure time	bar
Metering	cm^3
Coolant temperature	°C
Injection pressure limit	bar
Throat temperature	°C
Peripheral screw speed	m/s; m/min

9.1.2 Portability of Injection Speed Profile Switchover Points

When a mold is moved from one injection molding machine to another with a different screw diameter, all of the process' key positions with respect to the injection unit are altered and so they are not valid. The metering stroke (mm), the switchover point (mm), the injection speed profile switchover points (mm), will therefore be altered if a specific speed profile has been defined; these points will therefore have to be modified. It is necessary to modify all these values in order to adapt them to the machine injection unit's new dimensions. To avoid the trial-and-error method, we can calculate the conversion of these parameters. These calculated parameters will be very close to the optimum values.

Figure 9.2 is a spreadsheet for converting the filling time of an injection machine with one screw diameter to an injection machine with a different one and for converting the speed switchover points if a specific injection speed profile has been defined. To avoid the trial-and-error method, we can start with a conversion calculation of these parameters, which will give us the exact values which will be very close to the optimum values. These calculated parameters will be very close to the final values, taking into account the inertia, wear and calibrations of each machine, which probably need to be fine-tuned.

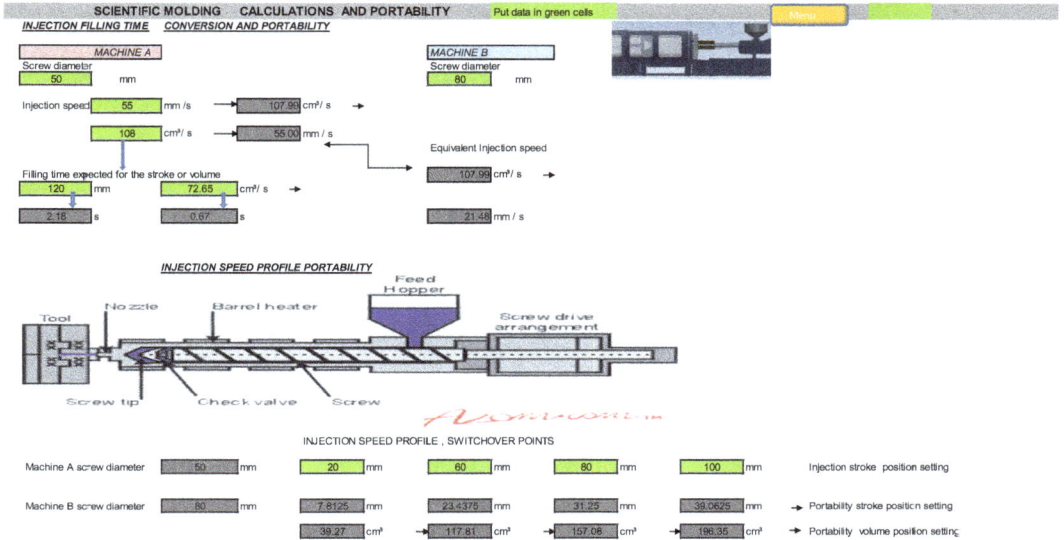

Figure 9.2 Example of a spreadsheet for the portability of the filling time and the injection speed profile switchover points from one injection machine to another with a different screw diameter

9.1.3 Portability of Injection Pressures

The injection unit in its hydraulic piston-screw assembly acts as a pressure multiplier. Figure 9.3 shows a synopsis of the piston-screw in an injection unit.

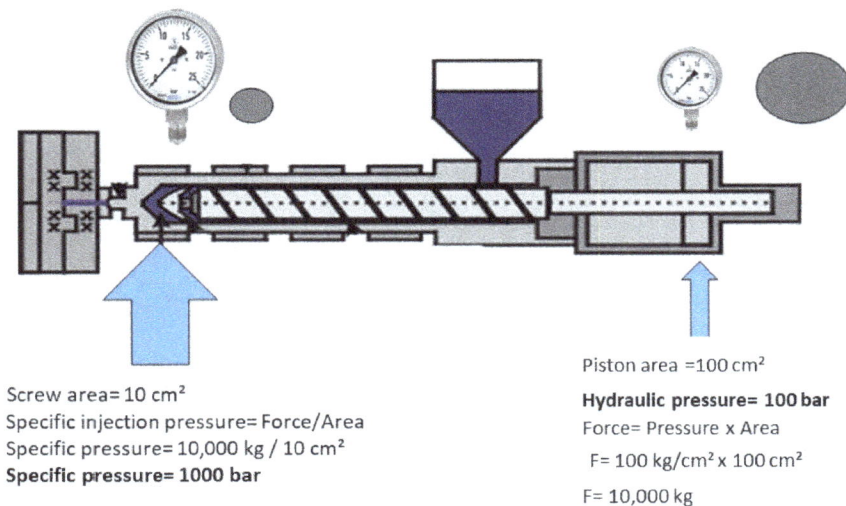

Screw area= 10 cm²
Specific injection pressure= Force/Area
Specific pressure= 10,000 kg / 10 cm²
Specific pressure= 1000 bar

Piston area =100 cm²

Hydraulic pressure= 100 bar
Force= Pressure x Area
 F= 100 kg/cm² x 100 cm²
F= 10,000 kg

Figure 9.3 Pressure multiplication diagram (hydraulic pressure is multiplied by 10 in this example)

Suppose a hydraulic pressure of 100 bar is applied to the rear chamber of the hydraulic injection piston. If the cross-section of the piston in the example is 100 cm^2, we will be exerting the following thrust force on the piston: 100 kg/cm^2 × 100 cm^2 = force of 10,000 kg. If this thrust force is applied at the other end of a screw section, like in the example in Figure 9.3 (introduced in Section 8.5) with a screw area of 10 cm^2, we will be exerting the following pressure on the material: 10,000 kg/10 cm^2 = 1000 kg/cm (approximately 1000 bar). We can conclude that a hydraulic pressure on the injection piston of 100 bar has been converted into a specific pressure on the material in the front area of the screw of 1000 bar. This example is for a multiplication factor of 1:10 (1 hydraulic bar equals 10 specific bar on the material). This multiplier effect is what the injection unit performs in each injection cycle.

We must be aware of this multiplication factor or intensification ratio of our machines if we want to have good portability; this will enable us to can convert hydraulic injection pressures into specific pressures on the material. This specific pressure value on the material is essential in scientific molding because, as mentioned, an advanced molder must focus more on the injection from a mold and the material standpoint rather than a machine standpoint. If we know the intensification ratio and therefore the specific pressure on the material, we can use this data to transfer the process from one injection molding machine to another with different characteristics. Nowadays, many modern machines already work with the specific pressure, which is perfect, but, in the case of working with hydraulic pressure, we need to know the multiplication factor.

Figure 9.4 shows how, with the same injection unit and the same hydraulic pressure, 100 bar, different specific injection pressures can be obtained as a function of screw diameter. This multiplication factor is also referred to as the intensification ratio; in the examples shown in red, intensification ratios of 1:7, 1:10, 1:12.5 can be obtained, with the values depending on the screw diameter in the same injection unit.

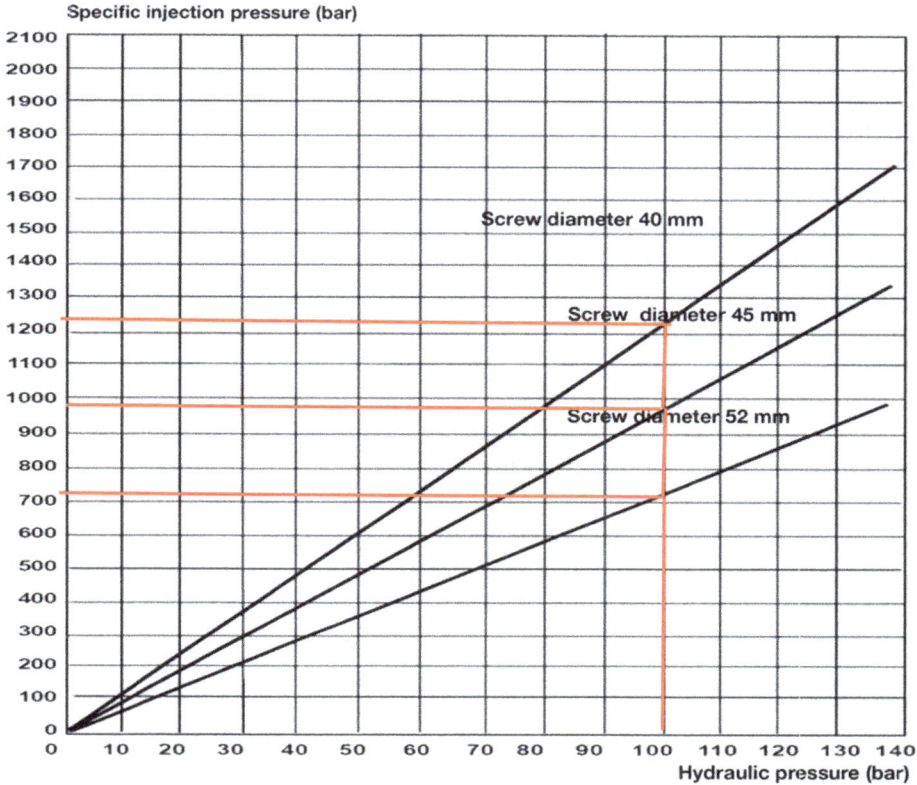

Figure 9.4 A graph showing the specific injection pressure with the same hydraulic pressure but different screw diameters

9.1.3.1 Why is it Important to Know the Intensification Ratio?

If we want to transfer a mold in production to another machine in our fleet using the parameters recorded in a parameter sheet where all the injection pressures are expressed in hydraulic values, we may have a problem when we want to reproduce the approved process on alternative machines.

Let us imagine that we have an approved process and that we are molding correct parts on an injection molding machine with a hydraulic injection pressure of 90 bar and an intensification ratio of 1:10. This would be equivalent to applying a specific pressure of 900 bar to the material. If we transfer this process to an injection molding machine with an intensification ratio of 1:7.5, this would mean that, with the same hydraulic pressure of 90 bar, we would have to apply a specific injection pressure of 675 bar to the material, much lower than the 900 bar approved for the correct process, and we would very likely obtain parts that are not acceptable, unfilled or that possess reduced mechanical and dimensional properties.

On the other hand, if we transfer the process to an injection machine with an intensification ratio of 1:12.5, this would mean that, with the same homologated hydraulic pressure of 90 bar, we would have to apply a specific pressure of 1125 bar to the material, much higher than the target pressure of 900 bar. Under these conditions, we will most likely obtain parts with flash, larger dimensions, heavier weight, etc.

Therefore, if we want to reproduce and carry out correct process portability from one machine to another with different characteristics, we must take into account the machine's multiplication factor or intensification ratio, amongst other things, in order to avoid making mistakes. This is one of the causes that explain why, when repeating the settings of similar machines or even machines believed to be identical but with a different intensification ratio, sometimes parts are obtained that are very different from those obtained in the original machine.

9.1.3.2 Hydraulic Pressure to Specific Pressure Conversions

We can calculate and convert the hydraulic pressure into specific pressure on the material if we know the intensification ratio (Figure 9.5). We can also calculate the equivalent hydraulic pressure in another injection machine with a different intensification ratio.

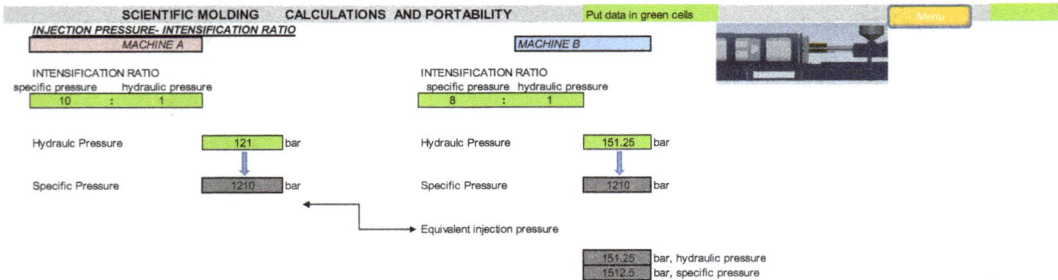

Figure 9.5 Example of a spreadsheet for calculating the intensification ratio and converting hydraulic pressure in specific injection pressure and the equivalent injection pressure for two different machines

9.1.4 Process Portability Format and Checklist

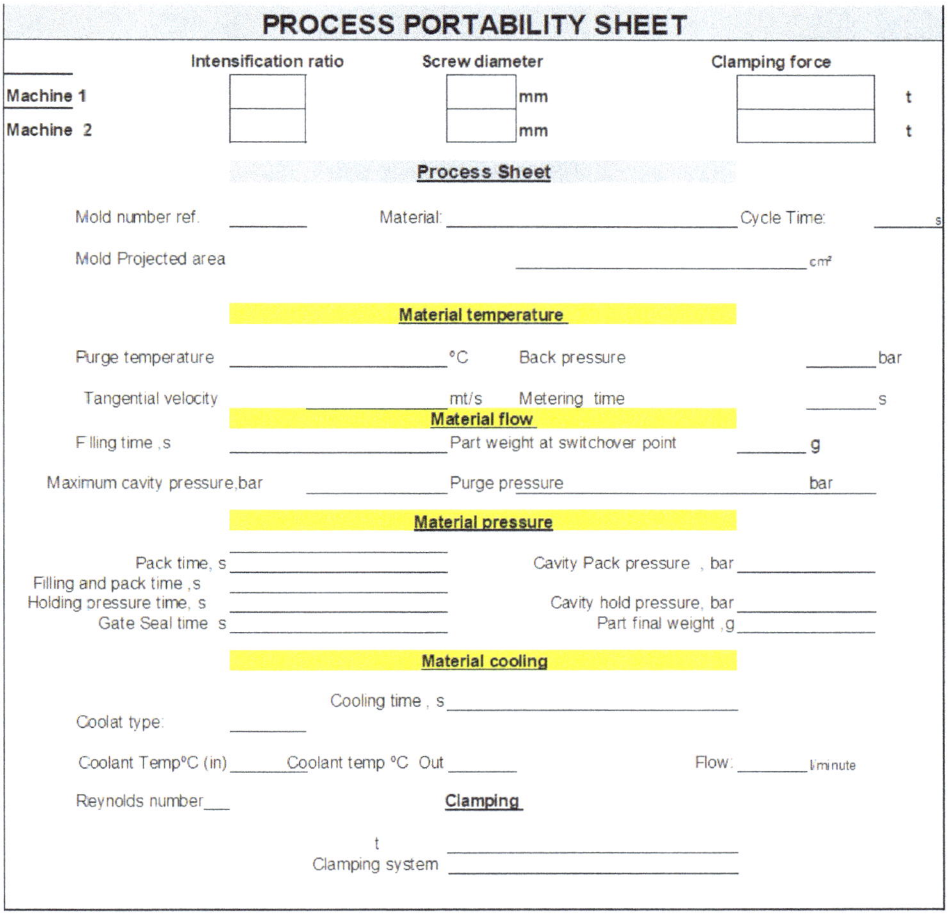

PROCESS PORTABILITY SHEET

	Intensification ratio	Screw diameter	Clamping force	
Machine 1		mm		t
Machine 2		mm		t

Process Sheet

Mold number ref. _____   Material: _____   Cycle Time: _____ s

Mold Projected area _____ cm²

Material temperature

Purge temperature _____ °C   Back pressure _____ bar

Tangential velocity _____ mt/s   Metering time _____ s

Material flow

F lling time ,s _____ Part weight at switchover point _____ g

Maximum cavity pressure,bar _____ Purge pressure _____ bar

Material pressure

Pack time, s _____   Cavity Pack pressure , bar _____
Filling and pack time ,s _____
Holding pressure time, s _____   Cavity hold pressure, bar _____
Gate Seal time s _____   Part final weight ,g _____

Material cooling

Cooling time , s _____
Coolat type: _____

Coolant Temp°C (in) _____ Coolant temp °C Out _____   Flow: _____ l/minute

Reynolds number___

Clamping

t _____
Clamping system _____

Figure 9.6 Example of a sheet for collecting all the information needed for good portability

PROCESS PORTABILITY		
Check list data for repeating an injection molding process in different injection molding machines		

Material temperature

	Barrel temperature setting for reach equal material purge temperature
	Specific back pressure, match
	Metering time, match
	Decompression, stroke and speed, match
	Peripheral screw speed, match

GOAL   **To match material thermal conditions**

Material flow

	Purge pressure, to match, design and diameter
	Filling time, match
	Part weight and switch-over point, match
	Maximum cavity pressure, match if sensors available

GOAL   **To match filling, hold pressure and volume injected**

Material pressure

	Hold pressure , match
	Final part weight , match
	Cavity fill, pack and hold time , match

If available mold sensors

	Cavity pack pessure, match
	Cavity hold pressure , match

GOAL   **To match cavity pressure , weight and dimmensions**

Material cooling

Coolant in and out temperature, match
Coolant flow, and coolant lay out conections, match
Cooling time, match

GOAL   **To match cooling time and cooling rate**

Figure 9.7 A checklist for repeating an injection molding process in different injection machines

Summary of Outputs for a Replicate Process

As summary, the following outputs should be recorded for fine portability of the process:

1. Melt preparation phase:
   a) Metering, peripheral screw speed
   b) Back pressure, specific pressure data

c) Decompression, speed and volume

d) Barrel melt temperature

e) Residence time

2. Filling phase:

a) Filling time, injection speed

b) Filling volume speed profile

c) Switchover volume

3. Hold phase:

a) Holding pressure, specific pressure data

b) Holding pressure time

9.2 Introduction to DOE (Design of Experiments)

Design of experiments (DOE) is a systematic, efficient method that allows technicians and engineers to study the relationship between various input variables and key output variables. It is a structured approach to collecting data and making discoveries. This type of DOE is used when there are several factors of a process or variables that are interrelated, and the effect of these factors on the outcome or response of the process is to be evaluated, as well as the effect of these interactions between factors on the outcome of the process.

9.2.1 Factors

These are process conditions that can be referred to as:

- Control: can be changed if necessary (e. g. mold temperature)

- Noise: cannot be controlled (e. g. material batch characteristics)

- Constant: do not change during the experiment or analysis (e. g. back pressure)

- Quantitative: can be changed in increments or decrements (e. g. holding pressure)

- Qualitative: can be changed in discrete levels (e. g. material batch or machine type)

In plastic injection molding, these factors are typically associated with data related to:

- Speed

- Pressure

- Time

- Temperature

- Volume

9.2.2 Responses

These are process outputs – the process' response to a certain combination of factors that cannot be directly controlled (e. g. filling time and filling pressure at the switchover point, cavity pressure, critical part dimensions, aesthetic defects on parts, etc.). These responses may be quantitative (e. g. a given part dimension) or they may be qualitative (e. g. aesthetic defects).

9.2.3 Levels

Number of values or magnitude of selected level of a factor, e. g. low holding pressure or high holding pressure, high melt temperature or low melt temperature, fast injection speed, medium injection speed or slow injection speed. Normally 2 levels are used most of the time in plastic injection molding.

9.2.4 Number of Experiments

The possible combinations of factors for each experiment are determined by their combination with the number of levels of each factor that we want to evaluate. Therefore, if we want to carry out an experiment with two factors, and the number of levels we want to study for each factor is two (e. g. low and high), the number of experiments would be four. This would be called a 2 × 2 factorial DOE. In this example, we would have to experiment with the combinations shown in Figure 9.8 in a factorial DOE (2 factors × 2 levels). The possible factors and levels combinations would be the four shown in Figure 9.9.

	Levels	
	Low	High
Factor Parameter 1	X	X
Factor Parameter 2	X	X

Figure 9.8 Experiments combinations for 2 parameters and 2 levels

	Levels		Experiment number
	Low	High	
Factor Parameter 1	X		
Factor Parameter 2	X		1
Factor Parameter 1	X		
Factor Parameter 2		X	2
Factor Parameter 1		X	
Factor Parameter 2	X		3
Factor Parameter 1		X	
Factor Parameter 2		X	4

Figure 9.9 Table matrix for experiments with 2 parameters and 2 levels

In other words, four experiments or combinations are necessary for a two-factor factorial DOE at two levels (2×2). If we were to study three factors instead of two, the number of experiments or combinations would be eight (2^3), and so on.

The equation for the number of experiments of combinations according to factors and levels is the number of levels raised to the power of the number of factors:

$$\text{Number of experiments} = \text{number of levels}^{nf}$$

where nf is the number of factors.

9.2.5 Experiment Matrix

An experiment matrix is a table showing information about the study:

- The input values for the experiment (factor, level, etc.)
- The outcome or response of each combination
- The number of experiments

The size of the experiment matrix depends on the number of factors and levels managed (seeTable 9.4). Normally two or three factors are used in plastic injection molding and two or three levels at most for a DOE.

Table 9.4 Number of Experiments as a Function of Different Levels and Factors

Number of experiments	Factors								
	1	2	3	4	5	6	7	8	9
Levels									
2	2	4	8	16	32	64	128	256	512
3	3	9	27	81	243	729	2187	6561	19683

Using the results obtained from a DOE, we can analyze the influence of different combinations of factors and levels in the process (see Figure 9.10 and Figure 9.11). By defining the key factors and levels, we can study a specific part of the process with these tools to analyze its robustness, critical points, etc.

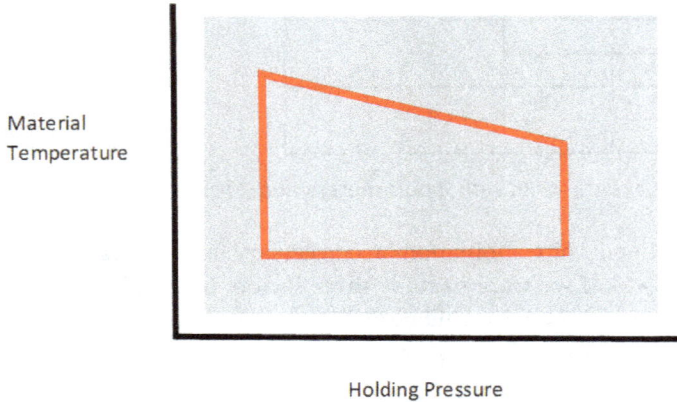

Figure 9.10
An example of a graph obtained from a DOE with two levels and two factors

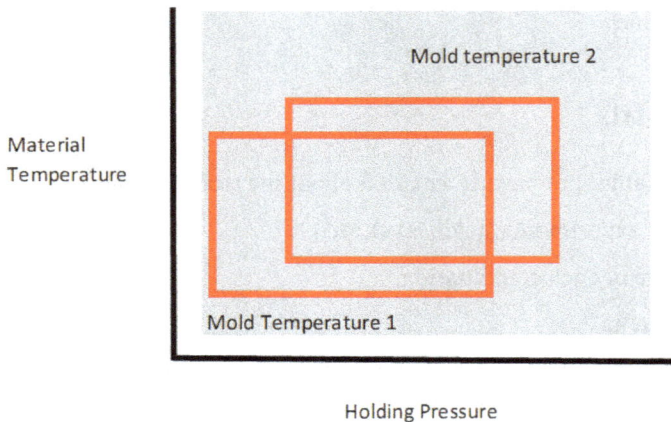

Figure 9.11
An example of a graph obtained with two levels and three factors

9.2.6 Injection Molding Factors or Parameters

In plastic injection molding DOE, the parameters or factors that could be used are those shown in Table 9.5.

Table 9.5 List of Factors or Parameters in Plastic Injection Molding

Number	Factor or Parameter
1	Injection unit temperature or melt temperature
2	Mold temperature
3	Injection filling speed
4	Injection pressure
5	Packing pressure
6	Packing pressure time
7	Holding pressure
8	Holding pressure time
9	Peripheral screw speed
10	Back pressure
11	Cooling time
12	Metering stroke or volume
13	Switchover point

All of these factors can be used as factors in injection molding studies or DOE. However, some of them could be highlighted as the most commonly used – for example, those listed in Table 9.6.

Table 9.6 Factor or Parameter Selection for DOE

Number	Factor or Parameter
1	Melt temperature
2	Filling time
3	Packing and holding pressure
4	Holding pressure time
5	Mold temperature
6	Cooling time

If, for example, we select three of these factors for a design of experiments and we apply two levels (high, low) there would be eight experiments to perform; from the calculation equation: $2^3 = 8$ experiments.

The most commonly used factors are shown in Table 9.6. These are considered in the next section.

9.2.7 Levels for Each Factor Selected

Once we have decided on the factors to be used in a DOE, the values of the levels to be used can be obtained from experiments or scientific molding tools instead of resorting to the empirical method.

9.2.7.1 Melt Temperature

We will choose within the polymer manufacturer's recommended range. Low values within this recommendation are used for thick walls and short flow paths; high values within this recommendation are used for thin walls and/or long flow paths. Low values are chosen for long residence time and high values for a short residence time.

9.2.7.2 Injection Speed (Filling Time)

The injection speed is selected within the range of values obtained from the relative viscosity curve or in-mold rheology test (see Sections 7.2.1 to 7.2.4). Values in the flatter portion of the curve can be selected as high and low levels for the injection speed factor. The mold filling time will then be a consequence of these selected values.

Once the two injection speed levels to be tested have been selected, it is very important to bear in mind that, when testing the higher injection speed value selected, we will probably have to modify the switchover point to ensure that we fill the cavity to the same percentage (95–98%) as for the lower speed selected. When the injection speed is increased, inertia is created that causes cavity filling to be higher at the higher speed, as well as the peak injection pressure in the cavity. If we want to determine the influence of each parameter in a factorial test, we must isolate other possible influences outside of these parameters by making small changes to the process. In this case, we would fill the mold with a different volume of material at the switchover point at a high injection speed rather than at low injection speed. We would be comparing different values.

9.2.7.3 Packing and Holding Pressure

The holding pressure levels are obtained by carrying out the process window definition test (Section 7.2.9), with the two temperatures selected for that experiment. We will obtain the minimum and maximum holding pressure values for the selected temperatures.

How to proceed:

Carry out a gate sealing time test (Section 7.2.8.3). Once the time from which the pressure is not applied in the cavity is known, when setting the holding pressure, start with the lowest pressure value, which will result in unacceptable parts being obtained. Increase the holding pressure level until acceptable parts are obtained, this will be the lower level to apply to this factor. Continue to gradually increase the holding pressure level until defects/problems arise (e. g. flash, ejection marks, or any symptom of over-pressurization), then slightly lower the pressure, which will reflect the maximum holding pressure level to be applied to this factor in the factorial experiment.

9.2.7.4 Holding Pressure Time

The holding pressure time will be the time during which we apply pressure to the cavity, where we will define dimensions, weights, surface finishes, possible stress, etc. In order to determine the levels to be used in the factorial experiment, we need to determine the gate sealing time (see Section 7.2.8.3) prior to the experiment. This gate sealing time will be the upper bound for this factor. Then we will select different holding-pressure times from the weight curve, more precisely on the vertical part of the shot weight curve.

9.2.7.5 Mold Temperature

Here, the values to be applied should be within the range recommended by the polymer manufacturer, considering that mold temperatures in the high recommended range should be chosen when we have thin walls, long flow paths, or we want to maximize crystallinity when molding semi-crystalline materials, with high gloss and appearance requirements, etc. Conversely, low temperatures within the recommended range should be chosen when we have thick walls, short flow paths, low crystallinity, more matte parts, etc. Mold temperature also has a direct influence on the cooling time and consequently on the cycle time.

9.2.7.6 Cooling Time

The cooling time mainly determines the ability to demold the parts from the cavity, the final dimensions of the part after shrinkage and the possible subsequent warping. To select this factor's levels, we can carry out the study in Section 7.3.3 to determine the influence of the cooling time in relation to a critical part dimension.

In summary:

- Melt temperature
- Normally levels recommended by the part manufacturer
- Injection speed (filling time)
- Obtained from the relative viscosity curve test, machine linearity and balance mold filling
- Packing and holding pressure
- Obtained from the process window test
- Holding pressure time

- Obtained from the gate sealing time test
- Mold temperature
- Obtained from the raw material manufacturer's recommendations
- Cooling time
- Obtained from the cooling time versus critical dimensions test

To apply these factors and levels, it is necessary to carry out the following tests before-hand:

- Relative viscosity test to select the two speeds to be applied
- Machine injection-speed linearity
- Mold filling balance
- Process window definition test to define the two holding pressure levels to be applied at the two selected temperatures
- Gate sealing time test to determine the two holding pressure times (they can be tested with a closed or open gate depending on the desired situation)
- Cooling time versus critical dimensions test to select the cooling times to be applied

9.2.8 Example of Experiment Matrix

Figure 9.14 shows is an experiment matrix for three factors and two levels. Number of experiments: $2^3 = 8$ experiments (2 levels and 3 factors).

Factors	Levels		
	Low	High	
Mold temperature	40	80	°C
Melt temperature	240	260	°C
Holding pressure	300	800	bar

Figure 9.12 Experiment data collected for three parameters and 2 levels

Number of experiments	Factors								
	1	2	3	4	5	6	7	8	9
Levels									
2	2	4	8	16	32	64	128	256	512
3	3	9	27	81	243	729	2187	6561	19683

Figure 9.13 Number of experiments table as a function of levels and factors or parameters

Example of Experiment Matrix

Experiment number	Mold temperature		Melt temperature		Holding pressure		Results Part dimension		
	Low	High	Low	High	Low	High	Length	Width	Height
	40	80	240	260	300	800			
1	40		240		300				
2	40		240			800			
3	40			260	300				
4	40			260		800			
5		80	240		300				
6		80	240			800			
7		80		260	300				
8		80		260		800			

Figure 9.14 An example of a three-factor, two-level experiment matrix for determining the best parameter combination for part dimensions

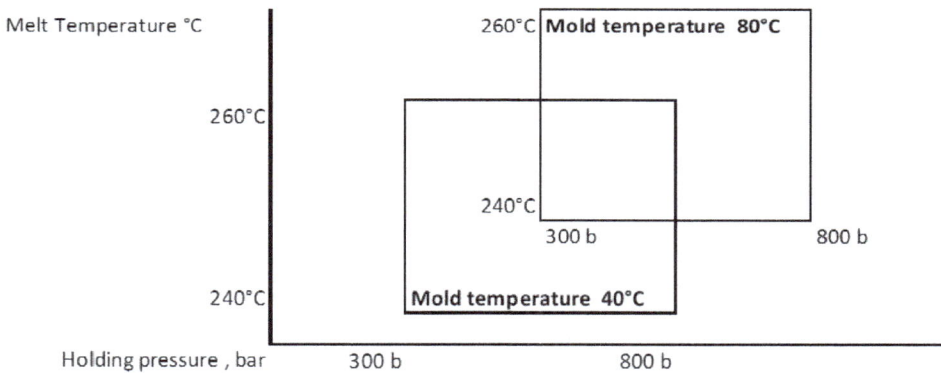

Figure 9.15 A graph corresponding to the experiment matrix

Through factorial experiments, we can sometimes better understand the relationship between inputs and outputs (i. e. between parameters and their results).

9.3 Consistent and Robust Process through Experimentation and Testing

A process that consistently produces parts is every injection molding technician's goal. There are three types of consistency:

- Consistency between shots: when the parts of each shot are identical, cycle after cycle

- Consistency between cavities: when the parts from different cavities are identical

- Consistency between series: when the parts produced series after series are identical

When a process is defined by achieving these three conditions, we can consider it to be a robust and consistent process.

The final quality of the molded parts depends on several factors, which could be grouped as follows:

- Part and mold design

- Material

- Mold

- Machine

- Injection process

Focusing on the injection process, this is a process with several variables and parameters, which in turn can be grouped into subfamilies:

- Temperature

- Time

- Speed

- Pressure

- Volume or position

This whole scenario of factors and parameters can be better managed through the application of scientific injection molding methods and with the use of DOE as a complementary tool. Scientific injection molding can be applied to the injection molding process, with some authors claiming it can be done in six steps or studies (discussed in Chapter 7) that are carried out in a specific chronological order:

1. In-mold rheology or relative viscosity study: used to determine the ideal injection speeds in terms of the material, which will allow us to inject with minimal influence from the shear rate on the viscosity of the material

2. Cavity filling balance study: used to ensure the balanced filling of all cavities, which means that we will have the same amount of material in the cavity when the packing and holding phase begins

3. Pressure loss study: used to determine the pressure losses along the flow path to ensure a sufficient injection pressure that will not limit the process. It is also used to determine at which stages of the flow path the pressure losses are excessive so that they can be minimized. Remember higher pressure losses means higher process variability

4. Gate sealing study: used to determine the time in which the gate will be permanently closed so that, on the one hand, we ensure that we will not have reflux or backflow and, on the other hand, we are not with excessive holding pressure time, after the gate is closed, the pressure and time applied doesn't affect the part dimensions, weight, shrinkage, etc. being therefore a waste of time and energy.

5. Cooling time study: used to determine the correlation between important dimensions of the parts and the cooling time so that the ideal cooling time can be defined

6. Process window study: used to determine the limits within which the process will produce acceptable parts, since the parts will be unacceptable outside of this molding window. Aesthetic or dimensional process windows can be used with different parameters, normally the melt temperature and the holding pressure

All of this is accompanied by the essential progressive filling study, the Delta P determination study, and the injection-speed linearity test.

The application strategy of these tools is to achieve the objective of defining an optimized, robust, and consistent process, not to obtain optimized part dimensions. Once the process is optimal (i. e. robust and repeatable), if the dimensions are not as expected, the next step is to adjust the mold to the required dimensions while maintaining the optimal process conditions. This way we have a robust process and acceptable parts.

9.4 Scientific Process Flowchart for Mold Qualification and Process Validation

This is an example of the application of the scientific injection molding tools in a logical order suggested by the author; see Figure 9.16. The user may employ their own order of application flowchart at their discretion and judgment, depending on the type of part, sector, requirements, etc., but always considering the safety of people and equipment, since the user is responsible for any possible consequences.

A defined ensemble of on-machine tests should be carried out through the application of the scientific molding tools, with the objective of defining a robust, consistent and optimized injection molding process on the one hand, and to check and validate the mold as an essential tool in a robust injection molding process on the other.

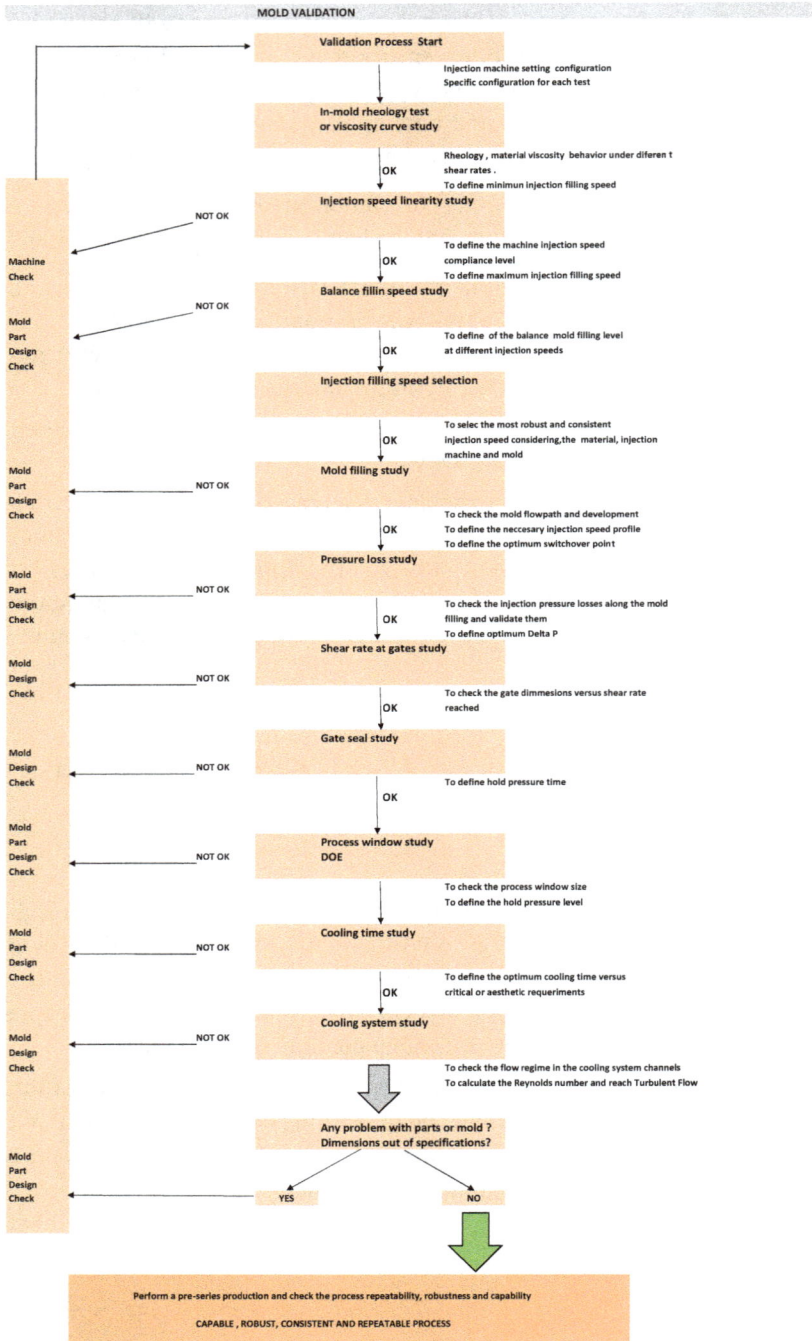

Figure 9.16 Example of a scientific process flowchart for mold qualification/process validation

It should be noted that the first step is to carry out the necessary tests to define a robust process using some of the scientific injection molding tools (i. e. defining the most robust mold filling speed, gates and runners, balance cavity filling, process windows, cooling study, cooling time, holding pressure, etc.). Once a robust and repeatable process has been defined, at the green arrow in Figure 9.16, we should ask ourselves the following question: If the process is studied and robust, do we obtain parts that comply with dimensional, aesthetic and functional specifications? If the answer to this question is affirmative, this would be the final point of the process definition procedure; the following step is to carry out a pre-series production run to check the long-term performance of the process. If the answer to this question is negative, it would be better to re-check the mold or part design to align dimensions and adjust functionalities while maintaining the robust process defined by the studies carried out.

10 Melt Preparation

The conditioning and preparation of the melt material is crucial to having a high-quality melt, due to the process conditions, as discussed in Section 8.11.1. This means that we will have a melt with the correct temperature, with a homogeneous temperature throughout the volume of the melt, with no unmelted material or degradation.

It is also crucial that the melt have an acceptable and constant density at the front of the screw once metering has been completed. As the material passes through the plasticizing unit, the cylinder and screw serve to melt and homogenize the melt by means of the barrel heaters and to compress and shear it by means of the rotation of the screw.

In addition, as the material is exposed to heat, its density decreases in proportion to the temperature, and so non-uniform temperatures will produce non-uniform densities (see Section 2.4.2).

10.1 Parameters that Influence Melt Quality

The following are some parameters that will help us to obtain a high-melt quality:

- Barrel temperature profile
- Throat temperature/tracking temperature
- Peripheral screw speed
- Back pressure
- Decompression stroke and speed

10.2 Barrel Temperature Profile

The barrel temperature profile is essential. This profile is one of the conditions that will define the real melt temperature and the machine's ability to produce, on the one hand, an acceptable melt quality and, on the other hand, a repeatable metering time, as well as to minimize the effect of a residence time which is either too long or too short.

As explained in Section 8.2, the real melt temperature output is the result of the following parameters or conditions: peripheral screw speed, back pressure, and injection unit temperature. To which must be added the compression generated by the screw design in its compression zone. To an extent depending on the screw design and the type of material, up to 80% of the heat required to melt the material can be provided here (see Section 2.4).

In any case, the melt temperature obtained must be within the processing range that is recommended by the polymer manufacturer and also directly dependent on material viscosity, mold design (gate, hot runner, runners, section, cooling system, etc.), and part design (flow path, length: thickness ratio, etc.). So, if we have to fill a mold with thin walls and long flow paths, we will choose a temperature in the high range recommended by the manufacturer, and conversely, if we have to fill a mold with thick walls or short flow paths, we will choose a temperature in the low range of the recommendations.

To minimize the effects of an inadequate residence time, the barrel temperature setting could have an increasing or decreasing slope, depending on whether the residence time is too long or too short. Figure 10.1 shows recommendations for three different barrel utilization percentages (80%, 50%, and 20%), with increasing and decreasing slopes, as well as temperature levels at three different residence times (3 minutes, 7 minutes, and 11 minutes). The idea is that, at high levels of percentage utilization, you should try to apply high temperatures to the material in the initial zones of the barrel, since the material will be in the screw for a short time. Conversely, at low levels of percentage utilization, you should try to apply high temperatures as late as possible in the front area of the barrel.

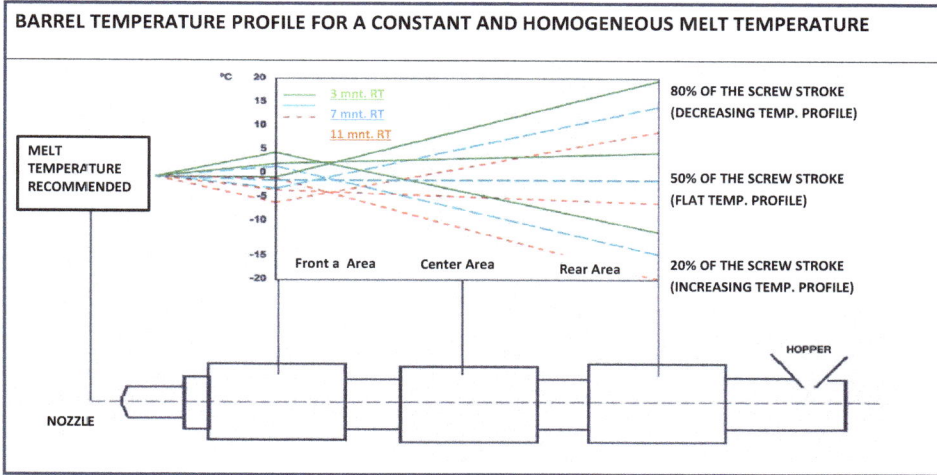

BARREL TEMPERATURE PROFILE FOR A CONSTANT AND HOMOGENEOUS MELT TEMPERATURE

Figure 10.1 Injection unit temperature profile according to barrel utilization (%) and residence time

10.3 Throat Temperature

This temperature is essential for a correct coefficient of friction between the material, barrel, and screw. The most important element of this coefficient is the temperature of the barrel next to the hopper in combination with the cooling system and the throat temperature, usually with water acting as a coolant. It is necessary to adjust this temperature so that we have a good "tracking temperature", which is the temperature that yields an adequate coefficient of friction (see Section 8.11.1.1). The goal is to find the temperature of the injection unit's rear zone (barrel and throat) that provides the shortest, most stable and repeatable metering time.

The tracking temperature can vary among different materials and provides the advantages of improving material filling inside the screw, preventing condensation in the throat, reducing the temperature difference between the first zone of the screw and barrel and the material throat below the hopper, improving the performance and durability of the rear heaters, and reducing slippage of the solid pellets with the screw by ensuring an adequate level of friction between the material and the inner surface of the barrel steel.

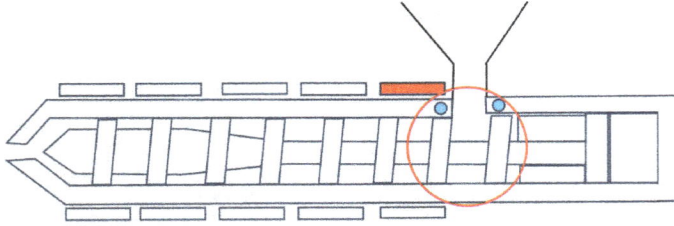

Figure 10.2 Throat area that must be thermally controlled for optimal processing

Table 10.1 Typical Throat Temperatures for Different Materials

Material	Material Name	Throat Temperature (°C)
HIPS	High-impact polystyrene	50–60
GPPS	General purpose polystyrene	40–60
SBC	Styrene-butadiene copolymer	40–50
ABS	Acrylonitrile butadiene styrene	50–60
PA 6 and PA 66	Polyamide 6, polyamide 66	70–90
PE	Polyethylene, low- and high-density	60–80
PPH and PPC	Polypropylene, homopolymer and copolymer	60–80
PEEK	Polyether ether ketone	80–95
POM-H and POM-C	Polyoxymethylene, homopolymer and copolymer	60–70
PPS	Polyphenylene sulfide	80–90
PMMA	Polymethyl methacrylate	50–60
PEI	Polyetherimide	80–90
PC	Polycarbonate	80–90
PC-ABS	Blend of polycarbonate and ABS	70–80
PC-PBT	Blend of polycarbonate and PBT	80–90
PPO-M	Modified polyphenylene oxide	70–80
PSU/PES/PPSU	Polysulfone, polyether sulfone, polyphenylsulfone	80–90
CA	Celulose acetate	40–60

Material	Material Name	Throat Temperature (°C)
TPU	Thermoplastic polyurethane	50–60
PET	Polyethylene terephthalate, semi-crystalline	60–70
PA11 and PA12	Polyamide 11 and polyamide 12	70–90
ASA	Acrylonitrile styrene acrylate	50–60
SAN	Styrene acrylonitrile	50–60

10.4 Peripheral Screw Speed

The rotation of the screw and its peripheral speed give rise to the shear acting on the material inside the injection unit (Figure 10.3). The shear rate reaches a maximum in the screw's external diameter zone, where the friction of the semi-solid or already molten material against the hot steel wall of the barrel generates the highest shear. These values can exceed the maximum permitted for each material; if we exceed them, the material will degrade, due to excessive shear rate, thereby losing molecular weight, properties and generating by-products as a result of extensive shear degradation.

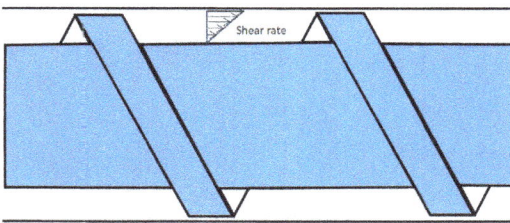

Figure 10.3 Shear rate inside the injection unit (higher values are reached at the external screw diameter)

The symptoms of excessive shear rate are usually bursts, tonality changes, brittleness, dark spots, etc. The maximum peripheral speed recommended by the polymer manufacturer should therefore not be exceeded (Table 10.2).

Table 10.2 Generic Minimum and Maximum Peripheral Screw Speeds Recommended by Polymer Manufacturers

Material	Minimum Peripheral Screw Speed (m/s)	Maximum Peripheral Screw Speed (m/s)
ABS	0.45	0.65
ASA	0.5	0.65
EVA	0.4	0.55
GPPS	0.7	95
PE-HD	0.65	0.8
HIPS	0.75	0.9
PE-LD, LPE-LD	0.6	0.75
PA11/12/6	0.3	0.5
PA6	0.3	0.5
PA66	0.3	0.5
PBT	0.25	0.35
PC	0.3	0.5
PC-ABS	0.35	0.55
PEEK	0.25	0.4
PEI	0.3	0.5
PETP	0.15	0.25
PES	0.15	0.25
PMMA	0.25	0.4
POM-H	0.1	0.3
POM-C	0.15	0.45
PP	0.65	0.8
PP-EPDM	0.45	0.65
PPO-M	0.3	0.5
PPS	0.15	0.3
PSU	0.125	0.2
SAN	0.3	0.45

10.5 Back Pressure

Back pressure is the effective pressure exerted by the material on the front of the screw during metering and metering steps (see Section 8.4). Applying the correct back pressure is essential for correct distribution and homogenization of pigments and additives in the material, correct thermal and melting homogeneity, and good melt plasticization. The back pressure is also very important for preventing the air trapped in the pellets from reaching the melt at the front of the screw.

10.5.1 Effects of Back Pressure

Advantages of correct back pressure include a more homogeneous melt, more frictional heat, displacement of air trapped in the pellets to the feed zone, reduced fluctuation in metering volume and cushion variability, and increased metered material density. The disadvantages that must also be taken into account are an increase in cycle time, due to longer metering time, and an increase in the heat generated inside the barrel.

The rotational movement of the "screw-barrel mechanism" in the compression and metering zones in each thread is essential for achieving homogeneity in the melt (see Section 2.4.2). Without this movement, the material, additives, colorants, etc. will not be well distributed, in terms of neither the melt temperature nor the density and will result in a low-quality melt and high variability in part quality.

To achieve correct melt homogeneity, it is essential to have enough back pressure at the front of the screw, have proper melt transport along the length of the screw, and have no wear within the screw and barrel diameter tolerances.

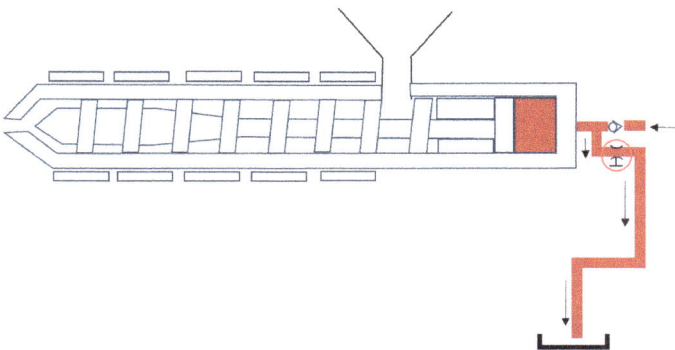

Figure 10.4 A diagram showing back pressure hydraulic control

10.5.2 Optimal Back Pressure Test

We can carry out a DOE to determine the most efficient back pressure. This will be the back pressure that gives us the smallest part weight variation in a series of sample shots.

To do this, perform ten injection shots with different back pressures and weigh the parts, calculating the weight range (Figure 10.5). A graph helps us to better understand the results (Figure 10.6), as also shown in Section 8.11.1.3.

Sample Shot	Back Pressure , bar						
	0	20	40	60	80	100	120
1							
2							
3							
4							
5							
6							
7							
8							
9							
10							
Maximum weight							
Minimum weight							
Weight range							

Figure 10.5

Table for collecting weight variation data at different back pressures

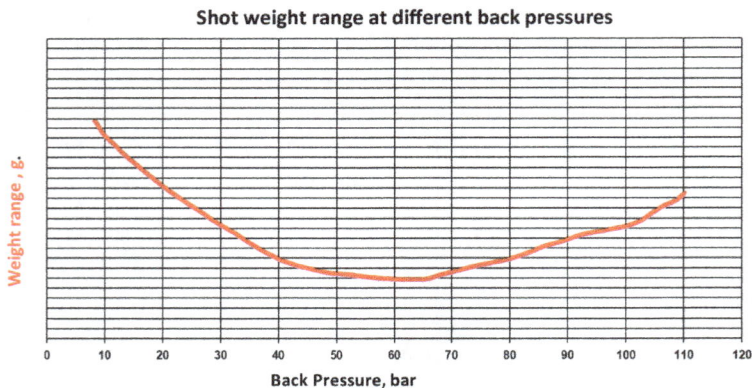

Shot weight range at different back pressures

Figure 10.6 A graph showing weight variation versus back pressure (here we can see that the back pressures between 50 bar and 70 bar give us the lower weight range)

10.6 Decompression or Suction

In current injection molding machines, we can set the decompression movement to occur before or after the metering movement. The pre-metering decompression movement is used to ensure that the screw tip seal ring is always in the forward position to maximize the volume of material that can pass through it during metering. This increases metering repeatability and is also very useful for decompressing the hot runners after the holding pressure phase.

The purpose of decompression, especially decompression after metering, is not only to relieve the pressure on the material in the front of the barrel, but also to bring the screw valve tip's sealing ring as far to the front as possible. This ensures that the closing stroke of this ring will remain constant and therefore the possible material leakage during the ring's closing movement will also be constant (see Section 8.11.1.2).

The speed of the decompression movement must be fast enough to ensure the correct position of the closing ring, but it must also enable the position reached by the screw, and therefore by the closing ring, to be repeated. In order to determine both the stroke and the speed of the decompression movement, a DOE can be carried out by controlling each cycle's total shot weight with different settings for both decompression stroke (Figure 10.7 and Figure 10.8) and speed (Figure 10.9 and Figure 10.10). The optimum setting will be the one that results in the smallest shot weight range.

Sample Shot	Decompression stroke (mm)						
	1	3	5	7	9	12	15
1							
2							
3							
4							
5							
6							
7							
8							
9							
10							
Maximum weight							
Minimum weight							
Weight range							

Figure 10.7
Table for collecting weight variation data as a function of decompression strokes

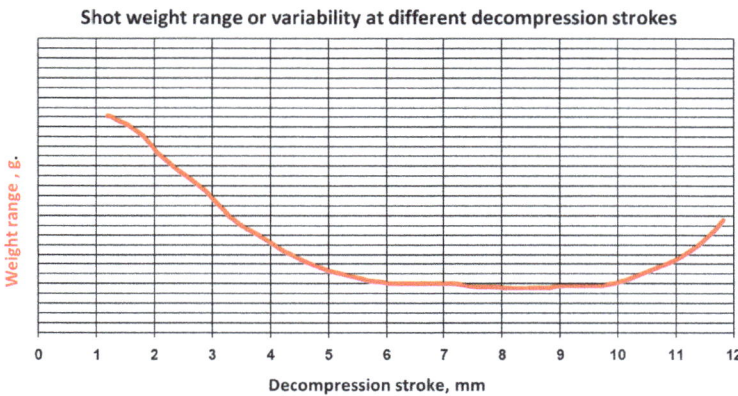

Shot weight range or variability at different decompression strokes

Figure 10.8 A graph showing weight variation as a function of decompression strokes

Sample Shot	Decompression speed (mm/sec)							
	10	20	30	40	50	60	70	80
1								
2								
3								
4								
5								
6								
7								
8								
9								
10								
Maximum weight								
Minimum weight								
Weight range								

Figure 10.9
Table for collecting weight variation data as a function of decompression speeds

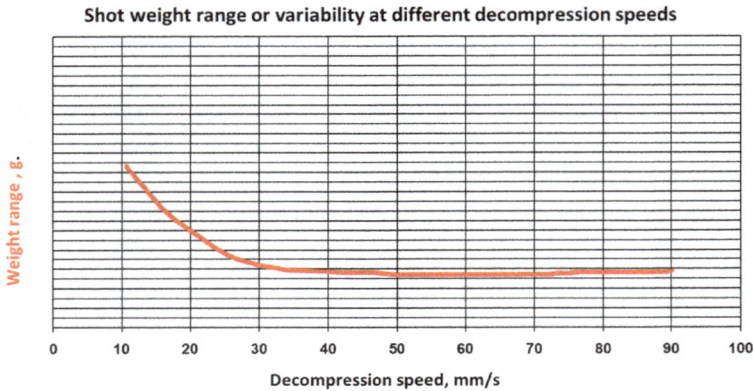

Shot weight range or variability at different decompression speeds

Figure 10.10 A graph showing weights or weight ranges versus decompression rates

The fine-tuning of the parameters described in this chapter yields a high-quality melt, which is crucial for the final quality of the parts and the repeatability of the injection process.

11 Process Variability, Self-Adaptation, and Corrections

11.1 The Plastic Injection Process, Self-Adaptation to the Process Variability

The injection molding machine and injection speed during dynamic mold filling:

The injection molding machine is designed and programmed to repeat the injection movement at an exact and repeatable speed and time. This means that the injection stroke from the metering position to the switchover point is performed at a repeatable and constant speed, even if the injection speed profile is flat or not.

In order to ensure this repeatability, the machine will use whatever injection pressure is necessary to achieve its set and target speed as long as injection pressure is available. Therefore, the machine should be able to adapt to different levels of force or load required to inject the material into the mold at the same set speed each and every cycle. This is the load sensitivity of each machine that can be tested (see Section 3.2.2).

This means that the machine will inject at a higher or lower injection pressure depending on variables, such as material viscosity, melt temperature, and mold temperature. In other words, the machine self-adapts to the desired injection pressure for the required load to achieve the injection speed and, consequently, the mold filling time.

These variables change the level of effort required of the injection molding machine to maintain and repeat the injection speed (and, therefore, the filling time) during the serial production of the molded parts. If the machine has pressure available, the injection speed must be repeatable, which will also make the mold filling time repeatable.

11.1.1 Mold Filling in the Dynamic Injection Phase

The various parameter settings that define and control the screw's forward movement during the filling of the mold, such as injection speed and speed profiles, are accurately reproduced in each cycle through the injection machine's adaptation process to the load required in each cycle; in this way, the machine continuously adapts the real injection pressure. In this phase, it is very important to have enough Delta P available for this pressure compensation or load adaptation (see Section 3.3).

During the dynamic mold filling phase, the machine repeatedly injects the same volume of material (the metering volume, from the metering position to the switchover point) into the mold in the same filling time, cycle after cycle, regardless of changes in the viscosity of the material. Therefore, if we do not cut the pressure curve by limiting the injection pressure and applying a correct Delta P value in this dynamic filling phase, the normal variability of material viscosity in different batches, recycled material, different moisture levels, etc., will not change the speed and conditions of filling the mold.

In this phase and under the aforementioned conditions, the machine will self-adapt to the viscosity, load changes and variability required.

11.1.2 The Filling, Packing and Holding Phase

During the holding phase, the most important settings are the holding pressure and holding time. This phase is what we can call a "pressure-limited phase" because the maximum pressure available to fill, packing and holding the material in the cavity is a fixed pressure value – the set holding pressure.

In this phase, the priority of the machine is not to maintain the injection speed, as was the case in the previous dynamic phase, where the machine self-adapts to the load or viscosity variation by applying the necessary injection pressure. Now the priority of the machine is to apply the set holding pressure for the set holding pressure time.

This holding pressure phase is therefore a "pressure-limited phase". In this phase, the injection molding machine will not be able to adapt to the variability of the viscosity and the process itself, nor will it be able to adapt to the holding pressure in each cycle depending on the load required by the viscosity variation.

At this stage, the viscosity variability will directly affect the pressure drops throughout the pressure transfer to the cavity, and therefore the pressure inside the cavity, leaving the machine with no way to self-adapt to these changes. This variability in pressure drop as a result of changes in viscosity – which cannot be compensated for by the machine in the holding pressure phase – affects the weight, dimensions and properties of the molded parts.

Therefore, if the viscosity increases, the pressure drops will be more significant and less pressure will be transferred to the inside of the mold. Conversely, if the viscosity decreases, the pressure drops will be smaller and more pressure will be transferred to the inside of the mold. These variations cannot be compensated for by the machine during the holding pressure phase with its limited set pressure.

11.2 Deviations from the Original Process Checks and Corrections

When a process variation is detected, such as a change in part dimensions, appearance, weight, and properties, the necessary parameters must be checked and corrected to return the process to the previous values or to return to the quality achieved by the original process.

11.2.1 Checks To Be Carried Out in the Dynamic Filling Phase

■ Real melt temperature

Check and compare that the melt temperature matches the recorded temperature. Variations in this value have a direct impact on the process. Parameters that affect this result include the peripheral screw speed, back pressure, barrel temperatures and residence time.

■ Injection pressure at switchover point

Check and compare the real specific pressure required to reach the switchover point with the recorded pressure from the original process. Higher pressures may indicate increased material viscosity and, conversely, lower pressures may indicate decreased material viscosity.

■ Filling time

Check the filling time and compare it with the original recorded time. These times should be equal. If not, check if there is enough Delta P available or if the process is pressure limited.

■ Switchover point

Check and compare that the switchover point has not changed. The volume injected in the dynamic phase should be the same (i.e. the weight of the parts at the switchover point should be the same as in the original process – if not, readjust the switchover point).

- Metering time

  Check the metering time and compare it with the recorded metering time. This time may be affected by key parameters, such as the screw speed, throat temperature, back pressure, decompression stroke and barrel temperature.

- Part weight at the switchover point

  Check the weight of the parts at the switchover point and compare it with the weights recorded from the original process. In the event of process deviations, it may be necessary to change and fine-tune the switchover point (and possibly the back pressure or the metering volume) in order to return to the initial weight at the switchover point.

11.2.2 Checks To Be Carried Out during the Holding Pressure Phase

Holding pressure

- Check the holding pressure setting and real pressure and compare them with the original recorded pressures.

- Compare the weight of the molded parts after this phase and compare it with the initially recorded values.

- Adjust the holding pressure to return to the initial weight values, increasing the holding pressure if the weight is below the standard weights and decreasing the holding pressure if the weight is excessive.

The use of this checklist and corrections to ensure that the weight of the molded parts at the switchover point and the weight after the holding pressure phase are equal to the original values recorded will ensure that the parts have the same dimensions, characteristics and properties as those initially produced.

These weight changes are often, but not always, caused by batch-to-batch variations in the viscosity of the material. Differences in the proportion and behavior of recycled material, differences in residual material moisture, differences in the colorants or masterbatches used, additives, and so on, can often cause variations in viscosity. On the other hand, dimensional differences can be caused by different additives, nucleation, mold temperatures, etc.

11.2.3 Differences to Check with Cavity Pressure Sensors

The separation distance between the two cavity pressure curves near the gate and at the end of filling equals the pressure loss necessary to fill the cavity (see Figure 11.1).

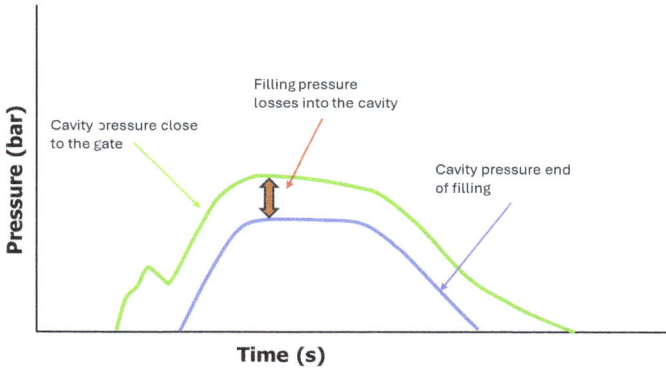

Figure 11.1 Graph showing hydraulic pressure curves and cavity curves near the gate and at the end of filling

Process differences with in-mold sensors

With in-mold pressure sensors, in the event of a viscosity change, e. g. change of material batch, the parameter adjustment would be:

- Adjust the switchover position so that the cavity pressure is the same as it was at the switchover point of the previous reference process.

- Adjust packing pressure so that the packing curve reaches the pressure level that of the previous reference process; this is the point of at which maximum cavity pressure is reached.

It may be necessary to increase the packing pressure to a higher cavity pressure level than we had with the previous batch in order to have a similar pressure at the end of the cavity filling. The objective is to reach a similar average cavity pressure (cavity pressure close to the gate + cavity pressure at the end of the filling/2) of the previous reference process.

11.2.4 Scientific Troubleshooter

Typical troubleshooters, using an empirical approach to the process, ask themselves, "Which button should I touch to eliminate the part defect or to center the process deviation?" They often use solutions based on troubleshooting guides.

The typical, empirical troubleshooter:

- Acts based on learned actions from past situations
- Makes multiple setting parameters changes at once
- Rarely documents changes in the process and process background
- Occasionally could damage equipment, mold, machine
- Avoids talking in specifics units

The **scientific troubleshooter**, with a scientific approach, asks:

- "What change has happened to cause the problem?" and looks at the process outputs
- "What is the physical root cause of the problem?"

Documentation of the process is crucial for these technicians, as everything regarding the process is important.

An advanced injection molding technician:

- Knows the history and background of the process, mold, machine, part, material and technology
- Determines what has changed: outputs, part, etc.
- Acts based on knowledge
- Verifies the result of each parameter change, waiting for the process to stabilize

11.2.5 Steps for Analyzing Deviations in the Injection Molding Process

Step 1

- Diagnosis and detailed analysis
- Collect all the relevant information regarding process, mold, part, machine and process background
- Check that there are no more defects present

Step 2

- Check that there are no basic root causes, where applicable (cooling system, material conditions, machine performance, etc.)

Step 3

- Compare with documented process information, outputs, and settings inputs
- Use process of elimination for the outputs and parameters unrelated to the problem or defect

Step 4

- Return the process to the documented standard outputs
- Change one parameter at a time
- With each parameter change, allow time for the process to stabilize

Step 5

- Verify and check all relevant parameters
- A complete review of the process may be necessary
- If the root cause is not detected, check the mold, machine, material, peripherals, etc.

Step 6

- Record actions, causes, and effects, troubleshooting details
- Ensure that other technicians can follow up all the recorded information

12 Data To Be Collected for the Calculation and Performance of a Scientific Injection Molding Process Methodology

The collecting of data, although extensive, is necessary for carrying out the studies, tests and calculations required for the correct development of a plastic injection process by means of scientific injection molding. This chapter summarizes all the information that needs to be collected in order to be able to use the calculation formulae, whether through spreadsheets or by directly applying the mathematical formula yourself.

The data are classified step by step according to the logical order of application in the development of an injection process following this methodology as that makes it easier to locate all data at each stage of the method. In the spreadsheets, the green cells are for data input for calculations, while the gray cells are calculation formulae or results from these calculations.

12.1 Data To Be Collected for Developing the Preliminary Studies and Calculations

It could be useful to make certain calculations before developing a process. Doing so provides us with prior information that can help us, for example, to select the ideal injection molding machine and to control some process limits that we are going to develop.

- Estimation of the theoretical clamping force required (Section 7.1.1)
    - Projected area of the part (cavity front area and runner system) (cm²)
    - Flow path (from the sprue to the cavity's farthest point) (mm)
    - Average part thickness (mm)
    - Injection pressure in the cavity (bar)

- Screw rotation, maximum rpm and peripheral screw speed (Section 7.1.2)
  - Screw diameter (mm)
  - Maximum peripheral screw speed (m/s; m/min)
  - Rpm applied (rpm)
- Metering stroke or volume calculation (Section 7.1.3)
  - Screw diameter (mm)
  - Part weight or part volume (g; cm^3)
  - Number of cavities
  - Density of the melt (g/cm^3)
- Theoretical cooling time (Section 7.1.4)
  - Average part thickness (mm)
  - Polymer processed
- Converting weight and volume
  - Part or shot weight (g)
  - Density of the melt (g/cm^3)
  - Polymer processed

12.2 Data To Be Collected for the Tests and Studies on the Reliability and Performance of Injection Molding Machines

Below is a summary of data to be collected for developing the test and studies to check the reliability and performance of an injection molding machine following the scientific injection molding methodology:

- Mold filling time repeatability study (Section 3.1)
  - Injection filling speed setting (mm/s)
  - Metering stroke or volume (mm; cm^3)
  - Screw diameter (mm)
  - Net dose (metering stroke minus switchover point position) (mm; cm^3)
  - Mold filling times reached at each speed tested (s)
- Load sensitivity study (Section 3.2)
  - Injection pressure at switchover point, with the melt and metering stroke reached (bar)
  - Filling time (s)

- Injection pressure at switchover point without load and melt, and the metering position reached by decompression (bar)
- Filling time (s)

- Delta P test (Section 3.3)
  - Injection pressure setting (bar)
  - Filling time (s)
  - Injection filling pressure reached (bar)

- Screw tip/check ring valve sealing dynamic test (Section 3.4)
  - Shot weight (g)

- Injection-speed linearity test (Section 3.5)
  - Metering stroke position (mm; cm^3)
  - Switchover point (mm; cm^3)
  - Injection filling speed (mm/s; cm^3/s)
  - Filling time (s)
  - Injection pressure (bar)

- Screw acceleration test (Section 3.6)
  - Metering stroke (mm)
  - Injection speed setting (mm/s)
  - Mold filling time (s)

- Pressure response test (Section 3.7)
  - P1: Pressure peak at the V/P switchover point (sometimes this is not the maximum injection pressure reached) (bar)
  - T1: Mold filling time during the dynamic injection stroke (s)
  - P2: Pack and holding pressure once the set pressure has been reached (bar)
  - T2: Real injection time elapsed when the holding pressure has been reached (s)

12.3 Data To Be Collected for Scientific Injection Molding Methodology Tests and Studies

- Viscosity curve test or in-mold rheology test (Section 7.2.1)
  - Metering stroke (mm; cm)
  - Switchover point (mm; cm^3)
  - Injection speed setting tested (mm/s)
  - Filling time (s)
  - Injection pressure reached (bar)

- Injection-speed linearity test (Section 7.2.2)
  - Metering stroke position (mm; cm^3)
  - Switchover point (mm; cm^3)
  - Injection filling speed (mm/s; cm^3/s)
  - Filling time (s)
  - Injection pressure (bar)
- Mold filling balance test (Section 7.2.3)
  - Injection filling speed setting (mm/s; cm^3/s)
  - Filling time (s)
  - Weight of each of the part's cavity (g)
- Analysis of injection pressure losses along the filling system (Section 7.2.6)
  - Maximum injection pressure available on the machine (bar)
  - Injection pressure required to purge (bar)
  - Injection pressure required to fill the sprue (bar)
  - Injection pressure required to fill the runner up to the gate, without passing through it (here we can have primary, secondary, tertiary, etc. channels) (bar)
  - Injection pressure required to pass through the gate (bar)
  - Injection pressure required to fill the cavity (bar)
- Delta P test (Section 7.2.7)
  - Injection pressure setting (bar)
  - Filling time (s)
  - Injection filling pressure reached (bar)
- Gate seal test (Section 7.2.8)
  - Injection holding pressure time (s)
  - Part weight (g)
- Process window test (Section 7.2.9)
  - Minimum temperature (°C)
  - Maximum temperature (°C)
  - Maximum holding pressure (bar)
  - Minimum holding pressure (bar)

12.4 Data To Be Collected for the Tests and Studies of Further Scientific Molding Tools or Checking and Improving the Injection Molding Process

- Shear rate at the gates study (Section 7.3.1)
  - For circular gates:
    - Part weight (g)
    - Polymer processed
    - Polymer density (g/cm^3)
    - Minimum filling time (s)
    - Step time between calculations (s)
    - Minimum gate diameter (mm)
    - Step diameter between calculations (mm)
  - For rectangular gates:
    - Gate depth (mm)
    - Depth/width ratio
- Cooling system study (Section 7.3.2)
  - Cooling system diameter (mm)
  - Coolant temperature (°C)
  - Flow through cooling system (l/min)
- Cooling time study, for a critical dimension (Section 7.3.3)
  - Cooling time setting (s)
  - Critical dimension (mm)
- Residence time study (Section 7.3.4)
  - Cycle time (s)
  - Screw diameter (mm)
  - Maximum metering stroke available on the machine (mm)
  - Useful metering stroke or net metering (metering stroke – decompression and cushion) (mm)
- Filling time repeatability study (Section 7.3.5)
  - Filling time (s)
- Screw tip or check ring sealing study – dynamic test (Section 7.3.6)
  - Shot weight (g)

12.5 Data To Be Collected for Process Portability Tests and Studies

- Portability of injection speed profile switchover points (Section 9.1.2)
    - Screw diameter (mm)
    - Injection filling speed (mm/s; cm^3/s)
    - Injection speed profile switch points (mm; cm^3)
- Injection pressure portability (Section 9.1.3)
    - Intensification ratio
    - Hydraulic injection pressure (bar)
- Screw rotation, maximum rpm and peripheral screw speed (Section 7.1.2)
    - Screw diameter (mm)
    - Maximum peripheral speed (m/s; m/min)
    - Rpm applied
- Metering stroke or volume calculation (Section 7.1.3)
    - Screw diameter (mm)
    - Part weight or part volume (g; cm^3)
    - Number of cavities
    - Density of the melt (g/cm^3)
- Converting weight and volume
    - Part or shot weight (g)
    - Melt density (g/cm^3)
    - Polymer processed

Summary of data needed to reproduce a process on different injection machines or for process portability:

- Data from the original injection process:
    - Screw diameter (mm)
    - Metering (mm; cm^3)
    - Shot weight (g)
    - Maximum metering stroke available (mm; cm^3)
    - Cycle time (s)
    - Filling time (s)
    - Intensification ratio
    - Injection filling pressure (bar)

- Back pressure (bar)
- Injection filling stroke – dynamic (mm; cm^3)
- Screw metering (rpm)
- Cooling time (s)
- Data from the alternative injection machine:
  - Screw diameter (mm)
  - Maximum metering stroke available (mm; cm^3)
  - Intensification ratio

12.6 General Table for Injection Molding Tool Selection According to Different Objectives and Situations

Given the large number of tools available within the scientific molding methodology, it is important to learn how to select the tools that we will use in each situation. In other words, we will not use all the tools in every situation (that would be a waste of time and effort), but we will scientifically select the right tools to obtain the information that will help us make decisions to define robust and consistent processes on a case-by-case basis. Figure 12.1 presents a summary table of the tools to be used in each situation.

	SCIENTIFIC INJECTION MOLDING TOOLS	Book Chaper	Preliminary Studies and Calculation	Injection Molding Machine Reliability and Performance
1	Estimation of the theoretical clamping force required	7.1.1 / 2.2.3	▓	
2	Injection pressure portability , intensification ratio	2.5 / 9.1.3	▓	
3	Screw rotation , maximum peripheral screw speed, and maximun rpm	7.1.2	▓	
4	Dosage stroke or volume calculation	7.1.3	▓	
5	Theoretical cooling time	7.1.4		
6	Viscosity curve or In-mold rheology	7.2.1		
7	Balance mold filling study	7.2.3		
8	Mold filling study	7.2.5		
9	Pressure loss study	7.2.6		
10	Gate seal study	7.2.8		
11	Process window DOE study	7.2.9		
12	Optimal injection speed tecnical selection	7.2.4		
13	Injecction speed linearity study	3.5--- 7.2.2		▓
14	Delta P test	3.3---7.2.7		▓
15	Dynamic and static screw tip sealing test	3.4---7.3.6		▓
16	Injection filling time repeatability test	3.1---7.3.5		▓
17	Residence time calculation	7.3.4		
18	Shear rate at the gate (circular gates)	7.3.1		
19	Shear rate at the gate (rectangular gates)	7.3.1		
20	Cooling system study	7.3.2		
21	Cooling time study for critical dimension	7.3.3		
22	Portability of injection speed and profile at switch positions	9.1.2		
23	Outputs and settings for a general process portability	9.1.1.2/9.1.1.3		
24	Load sensitivity test	3,2		▓
25	Screw acceleration test	3,6		▓
26	Pressure response test	3,7		▓
27	Optimal back pressure	10.5.2		
28	Decompresion or suction stroke and speed	10,6		

Figure 12.1 Tabular summary of all the scientific injection molding tools available and those which can be used in which situation. Gray cells indicate that that tool can be used for that situation or objective

Scientific Injection Molding Tools for Defining the Process	Further Scientific Injection Molding Tools for Checking the Process	Mold and Process Validation	Process Portability	Melt Preparation

13 Reference Data Tables

13.1 Maximum Residence Time

Material	Maximum Residence Time [min]	Temperature [°C]
ABS	5–6 / 2–3	265 / 280
ASA	5	270
PA 6	20	300
PBT	2 / 12	290 / 320
PC	7	320
PE	–	–
PET	4	290
PMMA	10 / 8	260 / 270
POM	7 / 20	240 / 210
PPO	10	280
PS	5	260
PS HI	5 / 1–2	260 / 280
SAN	5–6 / 2–3	265 / 280

13.2 Usual Mold Temperatures

Material	Mold Temperature [°C]	Material	Mold Temperature [°C]
ABS	60–90	POM	40–120
ABS/PC	70–100	PP	10–80
ASA	40–80	PPO	60–110
PA 6	60–90	PPS	150
PA 66	80–100	PPSU	150
PBT	80–100	PS	10–80
PC	80–120	PS HI	10–80
PE	20–50	SAN	40–80
PESU	140	TPC-ET	40–60
PET	130–140	TPE-V	10–80
PMMA	60–90	TPU	15–70

13.3 Shrinkage

Material	Shrinkage [%]	
	Flow Direction	Transversal Flow
ABS	0.5	0.5
ABS PC	0.6	0.6
ASA	0.3–0.8	
EVA	0.2–0.8	
LCP	0.07–0.5	
PA 6	0.9	
PA 66	1–1.6	
PA 66 GF 35	0.25	0.75–1.1
PBT	1.9	1.9
PBT GF 30	0.3	1.2
PC	0.5	0.7
PC GF 30	0.1	0.4
PC/PBT	0.7	1.1
PE-HD	2.4	
PE-LD	2.6	
PET	0.2	0.8
PMMA	0.2	0.5
POM	1.7–2.2	1.7–2.2
PP	1.2	1.3
PP T 20	0.93	1.05
PPO	1.2	1.6
PS	0.45–0.6	
SAN	0.5	

13.4 Drying Conditions

Material	Temperature [°C]	Time [hours]
ABS	80	4
ABS/PA	90	3
ABS/PC	110	4
ASA	80	4
LCP	150	4
PA 6/66	80	10
PBT	120/150	4
PC	120	4
PE	**	**
PET	135	5
PETG	65	6
PMMA	90	4
POM	85	3
PP 20% talc	80	3
PPO	80–100	2
PPO	110	3
PPS	140	4
PPS	150	6
PPSU	150	4
PS	70	3
PS HI	70	5
PSU	130	3
SAN	70	4
TPC-ET	100	3
TPU	90–100	4

** Dry is not neccesary

13.5 Maximum Allowed Moisture Data

Material	Starting Humidity [%]	Permissible Moisture [%]	Bulk Density [%]
ABS	0.4	< 0.2	0.63
ABS/PC	0.6	0.02	0.65
ABS/PA	0.6	< 0.1	0.65
ASA	0.35	< 0.1	0.63
LCP	0.04	0.01	0.97
PA 6	1	0.1	0.68
PA 6/30	0.6	0.1	0.94
PA 66	1	0.1	0.68
PBT	0.5	0.03	0.78
PC	0.1	0.02	0.75
PC OPTICAL	0.1	0.01	0.65
PET	0.2	0.004	0.85
PETG	0.3	0.07	0.8
PMMA	0.3	0.08	0.71
POM	0.8	0.1	0.85
PP	0.1	0.01	0.54
PP + Talc	0.2	0.03	0.7
PPS	0.1	0.01	0.81
PSU	0.3	0.05	0.75
TPU	0.4	0.02	0.72
SAN	0.3	0.2	0.65
SB	0.3	0.05	0.64

13.6 Recommended Depth of Venting Channels

Polymer	Easy Flow [mm]	High Viscosity / Glass Filled [mm]
ABS, SAN, HIPS	0.0508	0.0762
CA,CAB	0.0254	0.0381
PA 66	0.0127	
PA 66 15% FV	0.0127	0.0381
PBT	0.03	
PC	0.0381	0.0762
PE	0.0254	0.0508
PET	0.02	
PMMA	0.0508	0.0762
POM	0.0127	0.0381
PP	0.0254	0.0508
PPA	0.015	
PPO	0.0254	0.0762
PPS	0.0127	
PS GP	0.0254	0.0508
PSU	0.0254	
PVC	0.0254	0.0508
TPU	0.02	

13.7 Mold and Melt Temperatures, Shear, Etc.

Material	Name	Mold Temp. [°C]	Melt Temp. [°C]	Max. Melt Temp. [°C]	Max. Shear Speed × 1000 [s^{-1}]	Glass Temp. [°C]
ABS	Acrylonitrile Butadiene Styrene	60–90	200–260	280	30	100
ABS/PC	ABS-PC Blend	70–100	240–280	280	40	
EVA	Ethylene Vinyl Acetate Copolymer	20	140–220	220	30	
F PVC	Flexible Poly-vinyl Chloride	20	140–200	150	20	70
HIPS	High-Impact Polystyrene	20	200–260	280	40	
PA 6	Polyamide 6	80	230–280	320	60	40
PA 6 30% GF	Polyamide-6 30% Glass Fiber	80–90	270–290	320	40	
PA 612	Polyamide 6.12	80	230–280	320	60	
PA 66	Polyamide 6.6	80	270–320	360	60	50
PBT	Polybutylene Terephthalate	90	220–260	300	50	22
PC	Polycarbonate	80	280–320	320	40	152
PE-HD	High-Density Polyethylene	20	180–240	280	40	−95
PE-LD	Low-Density Polyethylene	20	180–240	280	40	−80
PET	Polyethylene Terephthalate	100–130	280–310	340	50	69
PMMA	Polymethyl Methacrylate	60–90	240–260	280	40	105
POM	Polyoxy-methylene	60–100	190–230	240	40	−38

Material	Name	Mold Temp. [°C]	Melt Temp. [°C]	Max. Melt Temp. [°C]	Max. Shear Speed × 1000 [s⁻¹]	Glass Temp. [°C]
PP	Polypropylene	20	200–280	300	100	−18
PPO	Polyphenylene Oxide	80	260–300	300	35	164
PPS	Polyphenylene Sulfide	100–150	310–240	360	50	
PS	Polystyrene	20	180–260	280	40	81
PSU	Polysulfone	150	330–400	420	50	
R PVC	Rigid Polyvinyl Chloride	20	140–200	200	20	
SAN	Styrene Acrylonitrile	60–80	220–260	280	40	
TPU	Polyurethane	20	190–220	260	40	

13.8 Maximum Peripheral Speeds

Material	Speed [m/s]	Material	Speed [m/s]
ABS	0.5	PE	0.8
ABS/PC	0.2	PET	0.3
ASA	0.3–0.6	PMMA	0.35
CA	0.45	POM	0.1–0.25
HIPS	0.5	PP	0.7
HYRTEL	0.4	PPA	0.2
PA	0.5	PPE/PA NORYL	0.4
PA 6	0.3	PPSU	0.6
PA 66	0.8	PS	0.7
PBT	0.35	SAN	0.55
PC	0.5	TPU	0.2

13.9 Density, Melt and Room Temperature

Material	Name	Melt Density [g/cm³]	Density at 23 °C [g/cm³]
ABS	Acrylonitrile Butadiene Styrene	0.92	1.05
ABS PC	ABS Polycarbonate	0.95	1.13
EVA	Ethylene Vinyl Acetate Copolymer		0.95
IONOMER	Ionomer		0.94–0.96
PA 6	Polyamide 6	0.96	1.13
PA 66	Polyamide 66	0.96	1.13
PA 66 GF 35	Polyamide 66 35% Glass Fiber	1.15	1.41
PBT	Polybutylene Terephthalate	1.07	1.3
PBT GF 30	Polybutylene Terephthalate 30% Glass Fiber	1.36	1.65
PC	Polycarbonate	1.04	1.2
PC / PBT	Polycarbonate/Polybutylene Terephthalate	1.04	1.22
PC GF 30	Polycarbonate 30% Glass Fiber	1.3	1.44
PE-HD	High-Density Polyethylene	0.72	0.957
PE-LD	Low-Density Polyethylene	0.7	0.917
PMMA	Polymethyl Methacrylate	1.04	1.19
POM	Polyoxymethylene	1.15	1.42
PP	Polypropylene	0.73	0.905
PP 40% FG	Polypropylene 40% Glass Fiber	0.85	
PP T 20	Polypropylene 20% Talc	0.87	1.04
PP T 40	Polypropylene 40% Talc	0.98	
PPO	Polyphenylene Oxide	1.06	1.1

Material	Name	Melt Density [g/cm^3]	Density at 23 °C [g/cm^3]
PPS	Polyphenylene Sulfide		1.65–1.95
PPSU	Polysulfone		1.29
PS	Polystyrene	0.92	1.05
PVC	Polyvinyl Chloride	1.15	1.43
SAN	Styrene Acrylonitrile	0.99	1.08
TPE V	TPE		0.92–0.98
TPU	Polyurethane		1.0–1.23

Bibliography

Michaeli, W., Greif, H., Kaufmann, H., Vossebürger, F.-J., Introducción a la tecnologia de los plásticos, 1992, Hanser, Munich

Arazo Urraca, J. L., Inyección de termplásticos, 2000, Plastic Comunicación, Spain

Hatch, B., On the Road with Bob Hatch, 1997, Injection Molding Magazine, Denver

Bichler, M., Seibold, G., Jäger, A., Rössner, F., Pahlke, S., La inyección en forma breve y sucinta, 2004, Demag Plastics Group, Germany

Tobin, W. J., Qualifications Start Ups and Tryouts of Injection Molds, 1992, WJT Associates staff, USA

Brydson, J. A., Plastics Materials, 1999, Butterworth Heinemann, Oxford

Harper, C. A., Modern Plastics Handbook, 2006, McGraw-Hill Education, Maryland

Berins, M. L., SPI Plastics Engineering Handbook, 1991, Springer, Massachusetts

Glossary

Anisotropy	The property of being directionally dependent. It can be defined as the difference, when measured along different axes, in a material's physical or mechanical properties. This property is found in some substances, especially those which are crystalline.
Back pressure	Pressure applied in the opposite direction to the advance of the material through the screw during the metering phase
Birefringence	Property of certain materials to split a light beam into two perpendicularly polarized beams
Butene	Olefin isomer (C_4H_8)
Calorific value	Amount of heat per 1 kg of substance released during combustion
Capillary rheometer	Instrument for measuring the rheological properties of plastics through temperature and deformation stress
Chemical bond	Cohesive forces exerted by pairs of electrons or ions that bind atoms together in a molecule
Chlorinated hydrocarbon	Volatile compound that is soluble in fatty tissues and used as a solvent
Closed-loop process	Self-regulating process
Coefficient of friction	Dimensionless coefficient indicating the slip resistance of a material
Commutation	See *Switchover point*

Comparative Tracking Index (CTI)	Maximum electrical voltage that a material resists before an electric arc is produced
Compression set	Loss of recovery of a specimen when compressed for a time at a given temperature
Covalent bond	Bond formed when two atoms share electrons to stabilize the link between them
Creep	Plastic deformation; also called cold flow
Crystalline	Description for a material formed by many very small crystals or crystallites
Crystallization	Formation of crystals or spherulites in the semi-crystalline polymer structure
Cushion	Melted material which remains in the front area of the screw after the holding pressure phase and ensures that the pressure is transmitted to the cavities
Delta P	Difference between the programmed injection pressure limit and the real injection pressure
Dew point	The dew point is the lowest temperature at which water vapor in the air begins to condense
Dielectric constant	Dielectric constant (Dk) characterizes the ability of a plastic to store electrical energy and can be defined as the ratio of the capacitance induced by two metallic plates with an insulator between them, to the capacitance of the same plates tested with air or vacuum as insulator between them
Dielectric strength	Dielectric strength reflects the electric strength of insulating materials at various power frequencies. It is a measure of the dielectric breakdown resistance under an applied voltage, and an indicator of how good a material is as an insulator. In plastics the typical measuring unit is kV/mm
Dissipation	Process by which friction is transformed into heat
DOE	Design of experiments is a systematic method for determining the relationships between the various factors affecting a process
Ductility	Ability to undergo deformation
Elastic modulus	Constant relationship between stress and elongation, within the elastic range of a substance

Exothermic reaction	Chemical reaction in which heat is released
FMEA	Failure modes and effects analysis; procedure for analyzing potential failures of a product
Free radical	Highly reactive molecule bearing an unpaired electron
Glass transition temperature	Temperature below which molecular motion is greatly restricted
Halogen	Element of group 17 (formerly VIIA) of the periodic table (bromine, fluorine, chlorine, iodine, astatine)
Heterogeneous	Compound or substance composed of two or more phases
Hexene	C_6H_{12}
Holding pressure	Pressure applied to the material after the dynamic filling phase to compensate for volume loss
Hydraulic injection pressure	Hydraulic pressure required for filling the cavities of a mold
Hydrolysis	Molecular breakdown caused by water and a consequent loss of properties
Hydrophilic	Having an affinity for water
Hydrophobic	Water-repelling
Intensification ratio	Ratio of hydraulic injection pressure to specific pressure acting on the material in the screw tip
Isochrones	In the case of a cavity-filling map, these show the advance of the material flow at equal time intervals
Isotropy	Uniformity in all orientations; when the properties are completely independent of direction
Jetting	Defect that occurs when molten material shoots into the cavity during filling and starts to solidify before the cavity is filled
K ratio	Screw compression ratio or the ratio of the fillet depth in the metering zone to the fillet depth in the plasticizing zone
Kinematics	The branch of classical mechanics which describes the motion of points, bodies (objects), and systems of bodies (groups of objects) without consideration of the causes of motion

L/D ratio	Ratio of screw length/diameter
Molecular weight	The sum of the atomic weights in the molecular formula of a compound
Molecule	The smallest unit retaining the properties of a chemical substance
Monomer	The structural unit from which macromolecules are formed
Melting temperature (T_m)	The temperature of melting of crystallites
Parting line	Parting line of a mold
Peripherals	Technical elements, attached to the injection molding machine for the manufacture of parts
Peripheral screw speed	Linear speed of screw during metering
Plasticizing capacity	Capacity of the injection machine if acting as an extruder
Polarity	Grouping of electrical charges within macromolecules
Polyaddition	Chemical reaction in which functional groups of monomers (or their ends) interact as a result of migration of a hydrogen atom
Polycondensation	Chemical reaction in which water or other substance of low molecular weight is secreted during the reaction
Polydispersity	Ratio of number-average molecular weight to weight-average molecular weight
PVT chart	Chart showing pressure, specific volume, and temperature
Reference sample	A reference sample approved by standards
Refraction	A change of direction undergone by a wave on passing from one material medium to another
Residence time	The time during which a material remains in the injection unit
Reticulation	The association of plastic molecules by chemical bonds to form a three-dimensional framework
Reynolds number	Dimensionless number used in fluid mechanics to indicate the flow rate
Scrap	Defective parts

Secondary attraction forces	Intermolecular forces whose range does not exceed a few nanometers
Setting	Data or parameters entered into the injector control system to define process conditions
Shear	Cutting stress between layers
Shear rate	Velocity gradient in a fluid between the different layers that compose it
Specific injection pressure	The effective pressure applied to the material for filling the cavities of a mold
Specific volume	Volume/mass ratio; the inverse of density
Spherulites	Small, rounded bodies produced by crystallization of the material; composed of multiple molecular chains
Sprue	A normally conical element that connects the mold inlet, where the machine contacts the mold, with the distribution channels
Surface tension	Molecular force at the surface of a body
Switchover point	Point at which dynamic pressure or cavity filling changes to static pressure or holding pressure
Throat	Initial area of the barrel under the hopper where the material inlet to the injection unit and screw is located
Viscoelastic	The behavior of a body that exhibits both elastic (Hooke's law) and viscous (Newtonian law) behavior
Viscosity	A measure of the resistance to gradual deformation by shear stress or tensile stress; the ratio of strain rate to applied stress
Vulcanization	Crosslinking or bonding between molecules

Index